T0358337

Synthetic Organic Chemistry and the Nobel Prize, Volume 3

The Nobel Prize is the highest award in science, as is the case with nonscience fields too, and it is, therefore, arguably the most internationally recognized award in the world. This unique set of volumes focuses on summarizing the Nobel Prize within organic chemistry, as well as the specializations within this specialty. Any reader researching the history of the field of organic chemistry will be interested in this work. Furthermore, it serves as an outstanding resource for providing a better understanding of the circumstances that led to these amazing discoveries and what has happened as a result, in the years since.

Series: Synthetic Organic Chemistry and the Nobel Prize

Synthetic Organic Chemistry and the Nobel Prize Volume 1
Dr. John D'Angelo

Synthetic Organic Chemistry and the Nobel Prize Volume 2
Dr. John D'Angelo

Synthetic Organic Chemistry and the Nobel Prize, Volume 3

John G. D'Angelo

CRC Press
Taylor & Francis Group
Boca Raton London New York

CRC Press is an imprint of the
Taylor & Francis Group, an informa business

First edition published 2025
by CRC Press
2385 NW Executive Center Drive, Suite 320, Boca Raton FL 33431

and by CRC Press
4 Park Square, Milton Park, Abingdon, Oxon, OX14 4RN

CRC Press is an imprint of Taylor & Francis Group, LLC

© 2025 Taylor & Francis Group, LLC

ISBN: 9780367438999 (hbk)
ISBN: 9781003006879 (ebk)

DOI: 10.1201/9781003006879

Typeset in Times
by Newgen Publishing UK

Access the Support Material: www.routledge.com/9780367438999

Contents

About the Author   vii

1   **Introduction**   1

2   **2001: Sharpless, Knowles, and Noyori**   53

3   **2005: Chauvin, Schrock, and Grubbs**   77

4   **2010: Heck, Negishi, and Suzuki**   96

Index   129

About the Author

John G. D'Angelo earned his BS in Chemistry from the State University of New York at Stony Brook in 2000. While at Stony Brook, he worked in Prof. Peter Tonge's lab on research toward elucidating the mechanism of action of FAS-II inhibitors for anti-mycobacterium tuberculosis drugs. He was also an active member of the chemistry club, serving as its treasurer for a year. After graduating, he worked as a summer research associate at Stony Brook in Prof. Nancy Goroff's lab, working toward the synthesis of molecular belts. He then earned his PhD from the University of Connecticut (UCONN) in 2005, working in the laboratories of Michael B. Smith. There, Dr. D'Angelo worked on the synthesis of 2-nucleobase, 5-hydroxymethyl lactams as putative antiHIV agents while also investigating the usefulness of the conducting polymer poly-(3,4-ethylinedioxy thiophene) as a chemical reagent. He served as a teaching assistant during most of his five years at UCONN and was awarded the Outstanding TA award during one of these years. After completing his PhD, he took a position as a post-doctoral research associate at The Johns Hopkins University in Prof. Gary H. Posner's lab. There, Dr. D'Angelo worked on the development of artemisinin derivatives as anti-malarial and anti-*Toxoplasma gondii* derivatives. In 2007, Dr. D'Angelo accepted a position at Alfred University at the rank of Assistant Professor, and in 2013, he was awarded tenure and promotion to the rank of Associate Professor at Alfred and awarded promotion to Professor in July 2021. Dr. John G. D'Angelo's research continued for a while to focus on the chemical reactivity of conducting polymers and has been expanded to pedagogical research and scientific ethics; the latter two now the focus of his research efforts. He served as the local ACS section (Corning) chair in 2014 and 2021 and as the Faculty Senate president for two consecutive terms serving in this capacity from 2014 to 2018 and became Chair of the Chemistry Division at Alfred in 2021. He is also the author of four books. One, on scientific misconduct, is in its second edition, and the second book on scientific misconduct is intended to be a workbook with hypothetical cases that students can work through. The third book, written with his PhD advisor, outlines a process for using the chemical search engine Reaxys to teach reactions, and the fourth book is a now discontinued organic chemistry textbook published through the web-based publisher Top Hat. He is also an author of 13 peer-reviewed publications (three in his independent career) and two patents. This four volume series on organic chemistry and the Nobel Prize is his latest authoring endeavor.

Photo credit:
My wife, Kerry Kautzman

Introduction

1

HISTORY

The Nobel Prize—a prize recognized at least in name—is one of if not THE premier reward for genius and is arguably the most famous award in the world. Nearly everyone past a high school education has likely heard of the Nobel Prize. Awarded (mostly) annually since 1901 in the subjects of Chemistry, Physics, Physiology or Medicine, Peace, and Literature and joined in 1968 by the Sveriges Riksbank Prize in Economic Sciences in Memory of Alfred Nobel, these prizes carry a medal, diploma, and cash prize for those chosen for this high honor in addition to the accompanying international recognition. It is important to note that the Economic Sciences prize is formally speaking not a Nobel Prize, though they're awarded at the same time and this prize is treated very much like a Nobel.

The prize was created by Alfred Nobel in 1895 in his last will and testament, with the largest share of his considerable fortune allocated to the series of prizes. The "rules" set out in Nobel's will about the award are still at least mostly adhered to today, over a century later. In his will, he specifically called for the Swedish Academy of Sciences to award both the Physics and Chemistry awards; the Karolinksa Institute in Stockholm to award the Physiology or Medicine award; the Peace prize to be awarded by a committee of five to be selected by the Norwegian Storting; and the Literature award to be awarded by the Academy in Stockholm. This protocol is still followed today. The chief difference is that although Nobel stipulated that the award should be made for a scientific matter from the *preceding year*, it appears (to me anyway) that it has more recently become more of a lifetime achievement award of sorts, at least in the life and physical sciences awards (Chemistry, Physiology or Medicine, and Physics). Exactly when this started is difficult to pin down, but it is very clearly the current modus operandi, even if only unofficially.

DOI: 10.1201/9781003006879-1

The primary source of Nobel's award-endowing wealth was as the inventor of dynamite, a stabilized form of the explosive nitroglycerine. In a very real way, the establishment of the prize was its own first controversy. Initially, the creation of the prize caused somewhat of a scandal, and it wasn't until several years after his death (1896) that his requests were finally fulfilled with the first series of awards (1901). His family opposed the prize; much of his considerable wealth had been bequeathed for its creation rather than to them so it is easy to understand their objection.

One may wonder why Alfred Nobel created the prize at all. Although there is no direct evidence to support the claims, legend has it that Alfred was horrified by an errant obituary—his own—mistakenly published upon the death of his *brother*. In it, Alfred was referred to as "the merchant of death," due to how his invention, dynamite, had been used. In addition to less violent uses (e.g., mining), dynamite and other modern derivatives/analogs are also used as a weapon. This much is an absolute matter of historical record. As for the legend, being labeled a "merchant of death" is enough to rattle just about anyone's emotional cage. The timing (1895) of his will (his third one) seems to fit this legend since the version of his last will and testament establishing the award was signed *seven years* after the death of his brother, rather than before his brother's death while his wealth had come far earlier.

Why Laureates?

Winners of the Nobel Prize are referred to as Nobel laureates, a reference with roots in ancient Greece. A laurel wreath, a circular crown made of branches and leaves of the bay laurel, was awarded to victors of athletic competitions and poetic meets in ancient Greece as a sign of honor. The term laureate to describe awardees of the Nobel Prize is used to harken back to this honor.

The Medal, Diploma, and Cash Prize

Nobel Laureates also receive a cash prize, an ornate diploma, and a medal. Created by Swedish and Norwegian artists and calligraphers each diploma is literally a work of art. The medals all have Nobel's image on them with his birth and death years, though the Economic Sciences prize—recall that this prize is in Nobel's memory and was not one of the original subjects endowed by Nobel—is slightly different in its design. Each medal is handmade out of 175–185 grams of 18-carat recycled gold. The cash award is currently set at nine million Swedish Krona. As of this writing, this is equal to ~851,000 Euro and ~965,000 USD.

OTHER FACTS

The awards are traditionally announced over a one-week period in early October. The ceremony and lectures then are held in Stockholm later in the year, usually in December. As of this writing, in mid-2024, since its inception, the Nobel Prize has been awarded to 962 individual Laureates and 30 organizations. Since great intellect and achievements often are not isolated, five individuals have won more than one Nobel Prize and three organizations have done likewise. Furthermore, some Nobel greatness apparently "runs" in some families. Among these, the Curie family is by far the most successful though they are not alone in having more than one family member earn an award.

Although there are no rules limiting how many awards one *recipient* can win, no award can be given for more than two works (i.e., topics) any given year. Moreover, the award may only be shared, whether it is for one work or two, between no more than three recipients among each of the awarded fields. In some cases, this sharing will happen by more than one topic earning the honor, other times, it will be due to more than one individual or organization contributing to the same work. In sum, the following permutations are the only ones allowed.

- One topic, one person
- One topic two people
- One topic three people
- Topic A one person and Topic B one person
- Topic A one person and Topic B two people

Nobel's will also stipulates what should happen in years where no award is given for a field. The will states, the prize money is to be reserved for the following year and if even then, no award can be made, the funds are added to the Foundation's restricted funds. This (no award for a field) has happened, between the awards, a total of 49 times, mostly during times of war. Technically, the statutes for skipping a year refer to it being possible to not make an award if "none of the works under consideration is found to be of the importance indicated in the first paragraph..." but that the award was consistently not awarded during times of global conflict is probably not a coincidence. It is unlikely that somehow, no important work was done during those years. Literature was skipped in 2018 amid controversy involving one of the committee members though in this case, it (along with the 2019 award) was awarded the next year, making delayed more accurate in this case than skipped. Peace is the one "skipped" the most at 19 times. The others do not even measure to half, in order: Physiology or Medicine, 9; Chemistry, 8; Literature, 7; and Physics, 6. The award for Economic Sciences has not existed during any times of global conflict and has never been skipped.

Prior to 1974, posthumous awarding of the prize was permitted and happened twice. Dag Hammarskjöl (1961, Peace) and Erik Axel Karlfeldt (1931, Literature) were both posthumously awarded the Nobel Prize. Both died earlier in the year they were awarded their prize making it likely they were, at the very least, nominated prior to their deaths. Hammarskjöl, the second Secretary-General of the United Nations was given the award "for developing the UN into an effective and constructive international organization, capable of giving life to the principles and aims expressed in the UN Charter." Hammarskjöl died in a plane crash in September 1961. Karlfeldt, meanwhile, was given the award for "the poetry of Erik Axel Karlfeldt" and died in April 1931. One exception since the 1974 moratorium has been made for extenuating circumstances. The committee did not know that the awardee, Ralph Steinman (2011, Physiology or Medicine) had passed away a mere three days before announcing the award. Steinman was given the award "for his discovery of the dendritic cell and its role in adaptive immunity."

The maximum number of recipients mandated by Nobel's will has become a modern source of controversy at least with respect to Chemistry, Physics, and Physiology or Medicine. Science, and in fact the world, is far more collaborative than it was in Nobel's time. Rarely—effectively never—does a modern researcher in these fields make a major or even small discovery alone. Instead, most modern scientific work involves literal scores of individuals playing an important, even if small, role in creating the final mosaic. Thus, restricting the Nobel Prize to such a small number of recipients guarantees that people contributing to the work are left out. It is even safe to say it leaves out *most* of the people contributing based on how modern science labs operate. That it is inevitably and arguably often the only people doing the lion's share of actual work being left out magnifies the problem. This disparity, or dare I say misappropriation of credit, is a fair criticism of the prize. However, we must recognize that it is also fair to lay this claim against *any* award. Even if it is more noticeable because of the magnitude of *this* prize, all awards (including sports awards) that recognize an *individual* ignore the contributions of essential supporting role players. By no means am I trying to justify or defend this reality. I only wish to point out here that all awards can be so criticized. This is covered in more detail, as are some potential resolutions, in the controversies section later in this introduction.

(Perhaps) surprisingly, two Nobel Laureates declined the prize: Jean-Paul Sartre (1964, Literature) on the grounds that he consistently declined *all* official honors and Le Duc Tho (1973, Peace) along with U.S. Secretary of State (who accepted the award), Henry Kissinger. Although they were jointly awarded the prize for their work on the Vietnam Peace Accord, Tho pointed to the ongoing situation in Vietnam as justification for declining. Four others were *forced* to decline the award. Three of the four were Germans—Richard Kuhn (1938, Chemistry) "for his work on carotenoids and vitamins," Adolf Butenandt (1939, Chemistry) "for his work on sex hormones," and Gerhard Domagk (1939, Physiology or Medicine) "for the discovery of the antibacterial effects of prontosil"—forbidden from accepting the award by Hitler; all three later were able to receive the diploma and medal but not the prize. The fourth, Boris Pasternak, a Russian, (1958, Literature) "for his important achievement both in contemporary lyrical poetry and the field of the great Russian epic tradition," initially accepted his award but was later coerced by authorities to decline it.

There are also no restrictions regarding the awardee being a free person; four laureates were imprisoned at the time of the award. Carl von Ossietzky (1935, Peace), Aung San Suu Kyi (1991, Peace), and Liu Xiaobo (2010, Peace) were all awarded the prize while incarcerated. von Ossietzky was given the award "for his burning love for freedom of thought and expression and his valuable contribution to the cause of peace" and was an anti-Nazi who revealed the rearmament efforts of Germany in violation of the Versailles Treaty that ended World War I. He was sent to a concentration camp when the Nazis seized power. Hitler's fury in response to von Ossietzky's award led him to prohibit all Germans from receiving the Nobel Prize (see Kuhn, Butenandt, and Domagk). Kyi was awarded the prize "for her nonviolent struggle for democracy and human rights." She opposed the military junta that ruled Burma, efforts that landed her under house arrest for nearly 15 years. After being released, she resumed her political career only to be arrested again after a military coup and later was sentenced to a total of eight years. Finally, Liu Xiaobo, given the award "for his long and non-violent struggle for fundamental human rights in China," received his sentence for the crime of speaking. His first stint in prison was due to his part in the student protests on Tiananmen Square in 1989 and a second (this time in a labor camp) for his criticism of China's one-party system. Most recently, in 2008, Liu co-authored Charta 08, which advocates for China's shift in the direction of democracy. His official charge was undermining the state authorities, and this earned him an 11-year sentence. In 2023, this club was joined by Narges Mohammadi. Narges was awarded the Nobel Peace Prize "for her fight against the oppression of women in Iran and her fight to promote human rights and freedom for all."

Some Nobel Laureates were downright deplorable, arguably more so than any of the aforementioned incarcerated Laureates. Take for example Dr. D. Carleton Gajdusek (1976, Physiology or Medicine), a pediatrician who discovered the role of prions in a disease known as Kuru, which is related to mad cow disease. He was also a self-admitted and I dare say unapologetic pedophile. Many of his victims were also his research patients. Another, Fritz Haber (1918, Chemistry) was potentially a war criminal. Both are covered in more detail in the controversies section of this introduction.

Sometimes, Nobel greatness runs in the family. A list of Nobel Prize-winning families is found in Table 1.1. The Curie family is the most prolific of the "Nobel Families." Pierre Curie won the prize (1903, Physics), sharing it with his wife, Marie (a.k.a. Madame) Curie, who went on to win one of her own (1911, Chemistry) several years later. One of their daughters, Irène

TABLE 1.1 List of Nobel Prize-winning families

CURIE			
Pierre and Marie Curie	1903	Physics	"in recognition of the extraordinary services they have rendered by their joint researches on the radiation phenomena discovered by Professor Henri Becquerel"
Marie Curie	1911	Chemistry	"in recognition of her services to the advancement of chemistry by the discovery of the elements radium and polonium, by the isolation of radium and the study of the nature and compounds of this remarkable element"
Irène Joliot-Curie and Frédèric Joliot	1935	Chemistry	"in recognition of their synthesis of new radioactive elements"
Henry R. Labouisse (on behalf of UNICEF)	1965	Peace	"for its effort to enhance solidarity between nations and reduce the difference between rich and poor states"
Cori			
Carl and Gerty Cori	1947	Physiology or Medicine	"for their discovery of the course of the catalytic conversion of glycogen"

TABLE 1.1   (Continued)

CURIE

Duflo and Banerjee

Esther Duflo and Abhijit Banerjee	2019	Economic Sciences	"for their experimental approach to alleviating global poverty"

Moser

May-Britt and Edvard I. Moser	2014	Physiology or Medicine	"for their discoveries of cells that constitute a positioning system in the brain"

Mydral

Gunner Mydral	1974	Economic Sciences	"for their pioneering work in the theory of money and economic fluctuations and for their penetrating analysis of the interdependence of economic, social and institutional phenomena"
Alva Mydral	1982	Peace	"for their work for disarmament and nuclear and weapon-free zones"

Bragg

Sir William and William Lawrence Bragg	1915	Physics	"for their services in the analysis of crystal structure by means of X-rays"

Bohr

Niels	1922	Physics	"for his services in the investigation of the structure of atoms and of the radiation emanating from them"
Aage	1975	Physics	"for the discovery of the connection between collective motion and particle motion in atomic nuclei and the development of the theory of the structure of the atomic nucleus based on this connection"

(continued)

TABLE 1.1 (Continued)

CURIE

Euler-Chelpin			
Hans von Euler-Chelpin	1929	Chemistry	"for their investigations on the fermentation of sugar and fermentative enzymes"
Ulf von Euler	1970	Physiology or Medicine	"for their discoveries concerning the humoral transmitters in the nerve terminals and the mechanism for their storage, release and inactivation"
Kornberg			
Arthur Kornberg	1959	Physiology or Medicine	"for their discovery of the mechanisms in the biological synthesis of ribonucleic acid and deoxyribonucleic acid"
Roger Kornberg	2006	Chemistry	"for his studies of the molecular basis of eukaryotic transcription"
Siegbahn			
Manne Siegbahn	1924	Physics	"for his discoveries and research in the field of X-ray spectroscopy"
Kai Siegbahn	1981	Physics	"for his contribution to the development of high-resolution electron spectroscopy"
Thomson			
J. J. Thomson	1906	Physics	"in recognition of the great merits of his theoretical and experimental investigations on the conduction of electricity by gases"
George Thomson	1937	Physics	"for their experimental discovery of the diffraction of electrons by crystals"
Tinbergen			
Jan Tinbergen	1969	Economic Sciences	"for having developed and applied dynamic models for the analysis of economic processes"

TABLE 1.1 (Continued)

CURIE

Nikolaas Tinbergen	1973	Physiology or Medicine	"for their discoveries concerning organization and elicitation of individual and social behaviour patterns"
Bergström and Pääbo			
Sune Bergström	1982	Physiology or Medicine	"for discoveries concerning prostaglandins and related biologically active substances"
Svante Pääbo	2022	Physiology or Medicine	"for discoveries concerning genomes of extinct hominins and human evolution"

Joliot-Curie and her husband Frédèric Joliot also went on to share a Nobel Prize (1935, Chemistry). This brings their family total to five shared or individual Nobel Prizes. As if this were not enough, although not an actual awardee, Henry R. Labouisse, husband of another of Marie and Pierre's daughters Ève, accepted the prize on behalf of UNICEF (1965, Peace). All told, this one family had a hand in no less than *six* Nobel Prizes. It is extremely unlikely that something like this will *ever* be matched. It appears that the Curies are the New York Yankees (with their all-sports record of 27 championships) of Nobel Prizes. The Curies, though the most prolific, do not hold an exclusive claim to multiple family members earning a Nobel Prize. Other spousal pairs to share a prize are: Carl and Gerty Cori (1947, Physiology or Medicine); Esther Duflo and Abhijit Banerjee (2019, Sveriges Riksbank Prize in Economic Sciences in Memory of Alfred Nobel); and May-Britt and Edvard I. Moser (2014, Physiology or Medicine) all shared a Nobel Prize while Gunnar Mydral (1974, Sveriges Riksbank Prize in Economic Sciences in Memory of Alfred Nobel) and his spouse Alva Mydral (1982, Peace) brought two separate Nobels to their house. Father and son pairs have also brought home Nobel Prizes with one pair: Sir William Henry and son William Lawrence Bragg (1915, Physics) sharing a prize. Other father-son pairs include Niels (1922, Physics) and Aage Bohr (1975, Physics); Hans von Euler-Chelpin (1929, Chemistry) and Ulf von Euler (1970, Physiology or Medicine); Arthur (1959, Physiology or Medicine) and Roger Kornberg (2006, Chemistry); Manne (1924, Physics) and Kai Siegbahn (1981, Physics); J. J. (1906, Physics) and George Thomson (1937, Physics); and Sune Bergström (1982, Physiology or Medicine) and son Svante Pääbo(2022, Physiology or Medicine). Rounding out the keeping it in the family trend are Jan (1969 Sveriges Riksbank Prize in Economic Sciences in Memory of Alfred Nobel), and younger brother Nikolaas Tinbergen (1973, Physiology or Medicine).

How Is the Nobel Prize Selected?

Each prize has a selection committee who, around September of the preceding year, sends confidential forms to individuals considered qualified and competent to nominate. Committee members are all members of the academy and serve a period of three years. Not just anyone can serve as an expert advisor; only those specially appointed.

Across the awards, nominations are not allowed to be revealed until 50 years after the prize has been awarded and the Nobel Prize website has a platform through which this can be searched. However, nominators are under no obligation to keep their nomination confidential. The qualified nominators and timeline (which does have some slight variations) for each award is summarized below.

CHEMISTRY

- Member of Royal Swedish Academy of Sciences
- Member of the Nobel Committee for chemistry or physics
- Nobel Laureate in chemistry or physics
- Permanent professor in the sciences of chemistry at the universities and institutes of technology of Sweden, Denmark, Finland, Iceland, and Norway and Karolinksa Instituet, Stockholm

- Holders of corresponding Chairs in at least six universities or university colleges selected by the Academy of Sciences with a view to ensuring appropriate distribution of the different countries and their centers of learning; and
- Other scientists from whom the academy may see fit to invite proposals.

PHYSICS

- Swedish and foreign members of the Royal Swedish Academy of Sciences
- Members of the Nobel Committee for Physics
- Nobel Prize laureates in physics
- Tenured professors in the physical sciences at the universities and institutes of technology of Sweden, Denmark, Finland, Iceland, and Norway, and Karolinska Instituet, Stockholm

- Holders of corresponding chairs in at least six universities or university colleges (normally, hundreds of universities) selected by the Academy of Sciences with a view to ensuring the appropriate distribution over the different countries and their seats of learning and
- Other scientists from whom the Academy may see fit to invite proposals.

PHYSIOLOGY OR MEDICINE

- Members of the Nobel Assembly at Karolinska Instituet, Stockholm
- Swedish and foreign members of the Medicine and Biology classes of the Royal Swedish Academy of Sciences
- Nobel Prize laureates in physiology or medicine and chemistry
- Members of the Nobel Committee not qualified under paragraph 1 above
- Holders of established posts as full professors at the faculties of medicine in Sweden and holders of similar posts at the faculties of medicine or similar institutions in Denmark, Finland, Iceland, and Norway

- Holders of similar posts at no fewer than six other faculties of medicine at universities around the world, selected by the Nobel Assembly, with a view to ensuring the appropriate distribution of the task among various countries
- Scientists whom the Nobel Assembly may otherwise see fit to approach
- No self-nominations are considered.

LITERATURE

- Members of the Swedish Academy and of other academies, institutions and societies that are similar to it in construction and purpose
- Professors of literature and of linguistics at universities and university colleges

- Previous Nobel Prize laureates in literature; and
- Presidents of those societies of authors that are representative of the literary production in their respective countries.

PEACE

- Members of national assemblies and national governments (cabinet members/ministers) of sovereign states as well as current heads of states
- Members of l'Institut de Droit International
- Members of the international board of the Women's International League for Peace and Freedom
- University professors, professors emeriti and associate professors of history, social sciences, law, philosophy, theology, and religion; university rectors and university directors (or their equivalents); directors of peace research institutes and foreign policy institutes

- Members of The International Court of Justice in The Hague and The Permanent Court of Arbitration in The Hague
- Persons who have been awarded the Nobel Peace Prize
- Members of the main board of directors or its equivalent of organizations that have been awarded the Nobel Peace Prize
- Current and former members of the Norwegian Nobel Committee (proposals by current members of the Committee to be submitted no later than at the first meeting of the Committee after 1 February); and
- Former advisers to the Norwegian Nobel Committee

ECONOMIC SCIENCES

- Swedish and foreign members of the Royal Swedish Academy of Sciences
- Members of the Prize Committee for the Sveriges Riksbank Prize in Economic Sciences in Memory of Alfred Nobel
- Persons who have been awarded the Sveriges Riksbank Prize in Economic Sciences in Memory of Alfred Nobel
- Permanent professors in relevant subjects at the universities and colleges in Sweden, Denmark, Finland, Iceland, and Norway
- Holders of corresponding chairs in at least six universities or colleges, selected for the relevant year by the Academy of Sciences with a view to ensuring the appropriate distribution between different countries and their seats of learning; and
- Other scientists from whom the Academy may see fit to invite proposals.

The last point effectively serves as an "out" of sorts to allow for solicitation from anyone in the field. Importantly, nobody can self-nominate. One would imagine (or hope) such hubris is unlikely to be received well by the committee anyway. Likewise, at least for the Chemistry prize it is not possible for *just anyone* to submit a nomination; for example, you cannot nominate one of your pals.

Chemistry,[1] Physics,[2] and Physiology/Medicine[3]

- September previous year: Nomination invitations sent out
- January 31: Deadline for nominations
- March-June: Consultation with experts
- June–August: Report writing with recommendations
- September: Academy receives final report
- October: Majority vote and announcement
- December: Ceremony

Literature[4]

- September previous year: Nomination invitations sent out
- January 31: Deadline for nominations
- April 15–20: Preliminary candidates
- May: Five final candidates
- June–August: Reading of productions
- September: Academy members confer
- October: Award announced
- December: Ceremony

Peace[5]

- September previous year: Nomination invitations sent out
- January 31: Deadline for nominations
- February–April: Preparation of short list
- April–August: Adviser review
- October: Majority vote and announcement
- December: Ceremony

Economic Sciences[6]

- September previous year: Nomination invitations sent out
- January 31: Deadline for nominations
- March–June: Consultation with experts
- June–August: Report and recommendation writing
- September: Academy receives report on finalists
- October: Majority vote and announcement
- December: Ceremony

Nobels Proven Wrong?

It is important to remember that science very routinely corrects itself. Without question, the mistakes being corrected by science are sometimes because of misconduct. Other times it is because technologies or other understandings improve, and we learn that the initial conclusions were simply wrong, despite the best contemporary efforts and science. It is very difficult to impossible to say which *really* happens more often, but I sense misconduct is the less common cause at least in part because misconduct, rather than a mistake or sloppy work, is not always easy to prove. It should be no surprise that work that earned someone a Nobel Prize is no different though it *is* rare. To date, no Nobel Prizes have been awarded for work later found to be fraudulent. Two examples of *disproven* works are Johannes Fibiger (1926, Physiology or Medicine) and Enrico Fermi (1938, Physics). Fibiger was awarded the Nobel for the discovery of a parasitic cancer-causing worm. Although the worm without question is real, subsequent research proved there were no cancer-causing properties of this parasitic infection. Fibiger died (1928) before his work could be proven wrong. Fermi, on the other hand, lived long enough to see his Nobel-winning work disproven and even agreed with the new (and correct) conclusions. Fermi had incorrectly concluded he generated new chemical elements during his nuclear chemistry work. He even went so far as to bestow names upon them. What Fermi had actually done, and Otto Hahn eventually was eventually awarded the Nobel for figuring it out (more on this later too), was cause nuclear fission to occur. That this phenomenon was never documented before without question led to Fermi's incorrect conclusion. It simply could not—at the time—been a logically arrived at conclusion. Only later did additional work shed light on what was *really* happening.

Other award-winning works seem dubious, at best, with modern hindsight. Paul Hermann Müller (1948, Physiology or Medicine) was awarded the Nobel Prize for his discovery of the anti-insect properties of DDT, a substance that however effective at controlling mosquitos—and as a result malaria suppression—is now banned globally except in very exceptional circumstances. Its environmental impacts were later found to be severe enough that DDT is one of the central topics in Rachel Carson's *Silent Spring*, a work credited with igniting the environmental movement. Carlson died relatively young at 57 (in 1964). Perhaps she would have been a Nobel Laureate (for Peace, I suspect) for her conservation work had she not died young. Finally, Antonio Egas Monic (1949, Physiology or Medicine) earned his award for developing the medical procedure lobotomy, a procedure rarely used today. Taken together, maybe these examples argue strongly in favor of the modern lifetime achievement approach. Such an approach allows the work to "marinate" or survive the vetting test of time. The flip side of this is that by then, the person may die, making it too late.

Nazis and the Disappearing Nobel Prizes

Something you are unlikely to find in your standard world history text is that two Nobel Prizes were dissolved to keep them from being captured by the Nazis as they overran Europe. German scientists, Max von Laue (1914, Physics), for his discovery of diffraction of X-rays by crystals; and James Franck (1925, Physics), with Gustav Hertz, for their discovery of the laws governing the impact of an electron on an atom, sent their Nobel medals out of Germany to fellow physicist Niels Bohr in Copenhagen for safekeeping. If caught doing this, von Laue and Franck likely would have been executed. Unfortunately, before long, Copenhagen was likewise now unsafe with the Nazis high-stepping through the streets. An associate of Bohr's, George de Hevesy, made the bold decision to hide the medals by *dissolving* them. That is right, he dissolved the gold medals much like any of us dissolve sugar in a cup coffee or a spot of tea. Gold is rather inert though, so dissolving it is difficult, requiring a solution that sounds like a bad idea to EVER make to the casual reader: a 3:1 mixture of hydrochloric acid and nitric acid. The ability of this mixture to dissolve gold (a noble-an unreactive-metal) is why it is referred to as aqua regia-kingly water. It oxidizes the gold, stripping away an electron to make the gold cation that is soluble in water. The result of this bold maneuver was two beakers of an orange solution. These beakers were set on a high shelf and de Hevesy eventually left the beakers behind as he fled for Sweden. As a Jewish scientist, he was not what you would call "safe" in Nazi-controlled land. After the defeat of the Nazis, De Hevesy retuned to the lab. Miraculously (or perhaps it was the ghost of Alfred Nobel or maybe just luck) the beakers remained untouched. A little bit of chemistry later, the gold cations were reduced and the pure gold precipitated out and was sent to Stockholm. The Swedish Academy had the medals recast, *from their gold*, and the medals were (re-)presented to their rightful owners in a 1952 ceremony. Personally, I find this story to be a wonderful thumb poked by science into the eye of arguably the most terrible scourge in human history.

OTHER MUSINGS

Donald Cram and the Wrong Number Phone Call

As silly as this may seem, even the Nobel committee can make a very amateur-seeming mistake.[7] In 1987, the Royal Swedish Academy of the Sciences called the wrong Donald Cram, calling Donald O. Cram, a carpet cleaner, instead of Donald J. Cram.[8] That the errant call woke up the entire household (6:10 AM local time) is only that much more humorous. This apparently happened one other time too, when the call was routed to a dentist, rather than the intended MIT professor.

Carolyn Bertozzi and Rage Against the Machine

In 2022, after the announcement of the Nobel Prize in Chemistry, one of the winners, Carolyn Bertozzi received public social media congratulations from what some may think is an unlikely source—lead guitarist for Rage Against the Machine, Tom Morello.[910] Why did Morello choose to single out Bertozzi, while staying silent on all other chemistry Nobel Prize awards? He and Bertozzi were bandmates at Harvard as undergraduates. I would be remiss if I do not take this opportunity to advocate for a diverse Liberal Arts education, through which one can meet and forge lifelong relationships with people in radically different backgrounds from your own. It also will provide an individual with a versatile background that will assist them in better putting their expertise and in turn their work in context with the rest of the world.

Why Is There No Nobel Prize in Mathematics?

It is curious, for sure, that the field of mathematics is omitted from the award Nobel endowed. One legend is that Nobel was bitter that his wife had an affair with a mathematician.[11] As Nobel was never married, this is unquestionably false. Even legends that replace the wife with a mistress do not hold water. Snopes in fact gives it a false rating.[12] More likely, he just didn't want to bestow an award for mathematics.

CONTROVERSIES AND SNUBS

Controversies

With honors such as the Nobel Prize, it should not be a surprise that there are occasional controversies, perceived snubs, or other general complaints and displeasure taken with both who is awarded the prize and who is not. There has even been some wondering whether the Nobel Prize is good for science at all.[13] Although a complete treatment of these complaints is not appropriate in this book series—it could likely be a book on its own—the interested reader is encouraged to perform a simple internet search for "Nobel Prize controversies." While the issue is far more commonly centered around the work or one of the awardees, entirely unrelated to the work of a Nobel Prize, in 2018, the Nobel Prize for Literature was delayed until 2019 because of sexual harassment claims made against one of the board members. These are not the controversies that will be discussed here though a brief discussion of some of the biggest work-based controversies is covered here.

There have been several Nobel Prizes that have been criticized in the sense that many believe the awardee should not have received the award. The Peace Prize seems particularly vulnerable to this. Former U.S. President Barak Obama (2009, Peace) is one of the most controversial in this respect. This is primarily because he was given the award a mere nine months into his presidency. Nominations are collected *months* ahead of the actual award. This means that (then) President Obama, given the award "for his extraordinary efforts to strengthen international diplomacy and cooperation between peoples" was nominated incredibly early in his presidency; almost upon becoming President. The notion that he, or frankly anyone, could have done enough to be worthy of this award is almost laughable.

Another controversial Peace Prize was given to Yasser Arafat, who shared the award with then Israeli Prime Minister Yitzhak Rabin and Israeli Foreign Minister Shimon Peres (1994, Peace) because of their collective work on the Oslo Peace Accords. Not only were these Peace Accords unsuccessful at fostering in peace—frankly calling into question all three of the awardees—Arafat was a key figure in many armed attacks in Israel; an awkward, at best, corollary.

Former U.S. Secretary of State Henry Kissinger's award (1973, Peace) was also very controversial. Recall, his Vietnamese counterpart Le Duc Tho refused to accept his share of the prize, citing that there was still conflict. There was also the bombing of Hanoi, ordered by Kissinger, during ceasefire negotiations but before the award was given. How this was not a disqualifying act on his part is difficult (at least for me) to understand.

Perhaps most scandalous of all is Fritz Haber (1918, Chemistry). Fritz Haber is the Haber of the Haber–Bosch process, a process still used largely unchanged today to make ammonia, essential to the commercial production of fertilizer. Because of this process, hundreds of millions, if not billions of people are fed. *This is not hyperbole.* Modern food production is possible because of his work. Whatever environmentalists want to (perhaps rightly) say about how sustainable this is and what the ceiling of productivity is, it is currently true. Just a few years *before* winning the prize, however, he orchestrated the first massive and deliberate chemical weapons attack during war, using chlorine gas against the allied forces in France during World War I.

D. Carleton Gajdusek (1976, Physiology or Medicine) is radically different from Fritz Haber. While Haber's crimes were done and well known before winning the prize, Gajdusek was not found to be a pedophile until years after his award. This means that Gajdusek's award was given ignorant of his vile actions while Haber was awarded his Nobel with full knowledge of the atrocities he committed. By no means am I comparing the crimes and ranking their "terribleness." There is, in my opinion, a stark difference between "holding your nose and making a decision" (Haber) and "later wincing at your decision" (Gajdusek) as disgusting personal information comes to light.

Nobel Snubs

Other controversies entail snubbing deserving winners. Before that, however, there are a number of giants of chemistry and science that could not have possibly been considered for the Nobel Prize, despite making contributions virtually unmatched in their importance. These pillars of chemistry, like those who solved the gas laws (Boyle, Charles, Gay-Lussac), Avogadro, and Lavoisier, all predated Alfred Nobel's founding of the prize. In fact, some predated Nobel himself. Lavoisier is arguably the most impactful among them. In addition to demonstrating the importance of oxygen to combustion—helping to finally vanquish phlogiston—his law of conservation of mass is perhaps the single most important contribution to chemistry. Thereafter, chemistry was a quantitative, rather than qualitative endeavor. It became a science according to more modern interpretations of science. Unfortunately for chemistry and especially Lavoisier, he was executed by guillotine at 50 during the French revolution. In the broader science sense, Charles Darwin died (much less violently than Lavoisier) in 1882, well before the creation of the Nobel Prize, rendering him ineligible. It should also be further noted that Darwin's work does not exactly fit into any of the areas recognized by the Nobel Prize. Physiology or Medicine is the closest match and would be quite the stretch. I encourage any reader wondering why "so-and-so" did not earn a Nobel to check the date of the person's death. Through this little bit of homework, you may discover they died before the Nobel Prize was created.

Such early scientists notwithstanding, it should come as no surprise—go ahead and be disappointed though—that the selection of the winner of the Nobel Prize is subject to the flaws of human behavior. By this I mean that inevitably, biases, discrimination, and personal conflict have at least appeared to play a role in denying some otherwise worthy work being overlooked.

Even divorced from various biases, every year, it can be argued that "someone else" should have won the Nobel Prize, and this is probably true in every field, not just chemistry—the overall focus of this series of books. In fact, this is the case for just about every award given across a very wide spectrum of recognition. Although sometimes the overlooked scientist eventually wins, the history of the Nobel Prize is littered with persons that it can be fairly asked "How did *they* never win?" Chemistry is not without such controversies. Also, multiple women have been snubbed a Nobel Prize in various fields. These specific snubs (of women) are covered in a later section of this introduction. The Nobel Prize website now also includes a nomination archive.[14] Although more modern nominations are still held secret, anyone interested can search for any researcher and find if they've at least been nominated during the years now publicly available. Some examples of general snubs are discussed below and when possible, I've indicated whether or not the person was at least nominated.

Gilbert Lewis, could have been awarded the Nobel Prize in Chemistry for more than one important discovery. Perhaps his most noteworthy contribution was the nature of a bond being the sharing of an electron pair. Lewis was nominated for the Nobel Prize many times between 1922 and 1946; more than half of the years over this time period in fact but never was awarded one.

Reasons for him not winning one are difficult to prove, but there is some evidence that personal conflicts or less than fully informed opinions on the part of some evaluators played a significant role.

Dimitri Mendeleev, father of the periodic table, was also never awarded a Nobel Prize in chemistry. He allegedly came close but the academy overseeing the award overruled the initial vote, changed the committee, subsequently holding a new vote that Mendeleev lost to Henri Moissan's benefit. Allegedly, this may have been at the behest of a rival of Mendeleev (who at the time was a member of the Swedish Academy).[15]

Henry Louis Le Châtelier was nominated for the prize but never awarded one despite his tremendously impactful contributions to chemical equilibria. It is possible that Le Châtelier's snub is a product of the historical fact that his most important work on equilibria was done too early (before the Nobel Prize was created), causing him to be overlooked because of the clause of "the previous year." According to the archive on the Nobel Prize website, Le Châtelier was nominated several times.

Chemistry and the sciences are not alone in being accused of snubbing deserving winners. Mahatma Gandhi, nominated five times for the Nobel Peace Prize, never won. Stephen Hawking, one of the most brilliant minds of our age was also never awarded a Nobel Prize for his work nor was Carl Sagan. This may have some root in the fact that Hawking's work at least could not be tested with contemporary tools, but maybe not. Both Hawking and Sagan are too recently active and deceased to make a confident conclusion on their nomination status.

George Washington Carver also very likely deserved a Nobel Prize. Carver was born into slavery but went on to become an agricultural scientist and promoter of environmentalism. Recognized by even the white community in a time still deeply steeped in terrible racial bigotry, he was dubbed a "Black Leonardo" (in and of itself a deeply problematic moniker in my opinion) by *Time Magazine* in 1941.[16] Carver's greatest contribution to the field was to improve types of soil depleted by repeated plantings of cotton. Carver died in 1943, which is important in the discussion of snubs since this is far longer than 50 years ago. Recall that nominations are sealed for 50 years, meaning that since no evidence can be found to support him being nominated, it appears he never was. Although agricultural science arguably can be under the umbrella of chemistry, that may be a more modern interpretation of the fields. This perhaps explains why he was never awarded one; however, there is no record of Carver ever being so much as nominated for a Nobel Prize. Considering he was never recognized as at least being worthy of nomination seems to be a colossal snub and it is at once hard to prove or ignore that racism may have played a role.

To be realistic and clear, the Nobel Prize is not the only prize awarded for excellence that snubs can be claimed against. The magnitude and worldwide recognition of the prize likely fuels the level of care people exercise over this particular prize.

WOMEN AND THE NOBEL

Over the years, across all the disciplines, the prize has been awarded to a woman 64 times, with one, Marie Curie, winning a total of two, in different fields (Physics with her husband Pierre in 1903 and Chemistry, alone, in 1911). The Physics prize has been awarded to women 5 times (2023, 2020, 2018, 1963, and 1903); Chemistry eight times (2022, 2020 (2 winners), 2018, 2009, 1964, 1935, and 1911); Physiology or Medicine 13 times (2023, 2015, 2014, 2009 (2 winners), 2008, 2004, 1995, 1998, 1986, 1983, 1977, and 1947); Literature 17 times (2022, 2020, 2018, 2015, 2013, 2009, 2007, 2004, 1996, 1993, 1991, 1966, 1945, 1938, 1928, 1926, and 1906); Peace, which leads the way with 19 times (2023, 2021, 2018, 2014, 2011 (three winners), 2004, 2003, 1997, 1992, 1991, 1982, 1979, 1976, 1946, 1931, and 1905); and the Economic Sciences prize has been awarded to a woman 3 times (2023, 2019, 2009). A numerical summary of these data are found in Table 1.2.

To date, all Nobel Laureates have outwardly presented as either male or female. Even a casual perusal of the data should make clear that there is a wild imbalance with respect to the gender of the Nobel Prize winners. Currently, there has never been a winner who has announced being trans and none of transitioned later in their lives after being awarded a Nobel Prize.

Though earlier on, the dearth of women recipients in especially the science fields could be explained away by a smaller number of women in the fields, compared to men, modern women win rates are nowhere near the averages reported by employers and (anecdotally, at least) observed by anyone paying attention to the world. It is difficult to pin down a plausible, data- or merit-based reason for this. Consequently, minds are left to wonder and inevitably wander to conclude discrimination and unfair judging. Although I want to say this is unjustified, I have no evidence to back up my stance. This is especially

TABLE 1.2 Analysis of percent of Nobel Prizes going to women

PRIZE	CHEMISTRY	PHYSIOLOGY AND MEDICINE	PHYSICS	PEACE	LITERATURE	ECONOMIC SCIENCES
Total	194	227	225	111	120	93
Women	8	13	5	19	17	3
% Women	4	5	2	18	14	2

true for the awards with a more closed nomination process. Until and unless there is better accountability (perhaps through more transparent nominations) or systematic data-based rubrics that generate nominations or awards, this pall is likely to hang over the prizes.

Commentary on the small number of women and person of color laureates, especially in the sciences, has been intensifying.[17] Although a shortage of nominees, at least for women has been cited,[18] concrete reasons for this lacking are far from agreed upon, though a shortage of women in the fields is demonstrably false given that employment data indicate that the percentage of women working in the fields is higher than the percentage of women taking home Nobel Prizes. Even considering the unofficial transition to more of a lifetime achievement award, rather than greatest achievement of the last year award does not hold water; women have been working in the sciences in large numbers for decades. A list[19] has been compiled of deserving women chemists (and there are likely lists for other fields as well) who arguably have been snubbed a Nobel Prize.

It would be quite easy to author an entire volume on a list of snubbed female Nobel-worthy scientists. The hardest part about such a list is determining when to stop. A short and nowhere near exhaustive list is given below.

Liese Meitner was one of the central contributors to nuclear fission, correcting Fermi's work along with Otto Hahn. Despite being a longtime collaborator of Hahn's, including on his award-winning work, Meitner did not share his award (1944, Chemistry). It is exceedingly difficult for anyone to justifiably rationalize this oversight even if some of her work was hampered as she fled for her life to Sweden (Meitner was Jewish) with the rise of the Nazis to power. Meitner was nominated several times for the Nobel Prize, including several times *with* Hahn. Curiously, 1944 was not one of these years though she was later nominated in 1946 and 1947.

Rosalind Franklin (one of the most infamous snubs) could fairly considered to actually be not snubbed in the sense that the award was given for the work she contributed to after her death. No sane and rational person can claim the committee waited for Franklin to die before recognizing the work. Nevertheless, it is impossible to ignore the fact that not one, not two, but three men were given the award related to her work. Part of what further enhances the ire of many is that particularly Watson and Crick collected little to no experimental results of their own regarding solving the structure of DNA, using especially Franklin's data on the way to do so. I for one believe—with admittedly no evidence—that she would have been awarded the prize had she not died. I see it happening two different ways. First, Franklin could have been awarded the prize instead of Wilkins. This in my opinion is the most likely course as her data was viewed to be of higher quality than his. A second option would have been to make two different awards whereby Watson and Crick would be given one (likely Physiology or Medicine though perhaps Chemistry) and the other pair would be given the Physics or Chemistry prize. There is no record of Franklin having been nominated for the Nobel Prize. Meanwhile, Watson, Crick, and Wilkins were nominated for the first time in 1960, several years *after* Franklin's death.

Eleanor Roosevelt was an ardent advocate for civil rights and was never awarded the Peace prize for her brave and noble work. Other civil rights activists (e.g., Rev. Dr. Martin Luther King, Jr.) were awarded the prize so a claim that civil rights advocacy is insufficient to earn the award is weak, at best. Eleanor Roosvelt was nominated for the Nobel Prize several times, as were her husband and of course Teddy Roosevelt (who received it in 1906).

Joselyn Bell Burnell, who performed the work that led to the discovery of pulsars sat by and watched while her colleagues won the Nobel Prize (1974, Physics). At the time the work was done, Burnell was a postdoctoral research student, further highlighting the disparity in credit between the Principal Investigators (PI) and those who work under their tutelage or mentoring. According to the archive, Burnell was never nominated for the prize.

Rachel Carlson's work is one of, if not *the* cornerstone in the environmental movement. Had she not passed away in 1964, she likely would have been awarded a Nobel Prize for Peace for her impact on the environmental movement. Instead, it wasn't until 43 years later that Al Gore and the UN's Intergovernmental Panel on Climate Change were awarded the Nobel Prize for Peace for their work on increasing awareness regarding climate change. According to the nomination archive, Carlson was never nominated for the prize.

RACE AND THE NOBEL PRIZE

The lack of racial diversity, particularly as it is measured by skin color is as shocking as it is disappointing. Consider Table 1.3, which summarizes the data.

In most of the prizes, the numbers of non-Caucasian awardees are approximately the same level as women. Here, however, it may be easier to rationalize but I want to fall well short of *justifying* it. This rationalization comes from the reality that at least in the cases of the research-based awards, the vast majority—in fact nearly all the winners—are affiliated with institutions within affluent countries who have sometimes billions of dollars annually invested in basic research. This higher level of funding inevitably leads to higher profile and higher quality—or at least higher sophistication— research. This is not to say that non-Caucasian researchers are for certain never overlooked. Once again, it appears inevitable that a black or brown person will be working at a high-profile and well-funded institution with vibrant research support. I even know more than a few such persons. As with women, there once again appears to be room to conclude that only discrimination can produce this reality.

TABLE 1.3 Numerical summary of race and the Nobel Prize

PRIZE	CHEMISTRY	PHYSIOLOGY OR MEDICINE	PHYSICS	PEACE	LITERATURE	ECONOMIC SCIENCES
Total prizes	194	227	225	111	120	93
# Caucasian winners	177	219	201	81	104	90
% Caucasian	92	96	89	74	87	97
# Birth countries	38	38	35	48	49	19
# Affiliated countries	20	24	21	45	35	10

It is likely that an entire book can be written about the issue of general equality (be it gender, racial, or any other kind) in not just the Nobel Prizes but all manner of awards. One take that I heartily disagree with is the claim that it is a reflection of racism in the American education system.[20] Although the American education system is certainly rife with *inequality* that has many diverse roots, including but not only racism, to claim that the American education system is at fault for a lack of diversity in a decidedly *international* award is absolutely off base maybe even tone-deaf and delusionally U.S.-centric. Though this *may* be part of why there is a dearth of racially diverse U.S.-based Nobel winners, it cannot possibly explain the issue on a global level. In fact, isolating the explanation to the education system in any country is potentially more counterproductive than it is helpful.

Another way of considering the diversity in awards can be by looking at nationalities. Outlined herein is a breakdown of country of origin and affiliation for all winners of each Nobel Prize category.

CHEMISTRY

By affiliation

U.S., 87	Israel, 4	Finland, 1
Germany, 33	Canada, 3	Italy, 1
U.K., 29	Argentina, 1	Netherlands, 1
France, 9	Austria, 1	Norway, 1
Switzerland, 6	Belgium & U.S., 1	Switzerland & U.S., 1
Japan, 6	Czechoslovakia, 1	U.S.S.R., 1
Sweden, 5	Denmark, 2	

CHEMISTRY

By birth

U.S., 58	Scotland, 3	India, 1
U.K., 25	Switzerland, 3	Italy, 1
Germany, 24	British Mandate of	Korea, 1
France, 12	Palestine, 2	Lithuania, 1
Japan, 7	Denmark, 2	Mexico, 1
Austria-Hungary, 6	Egypt, 2	New Zealand, 1
Prussia, 5	Hungary, 2	Poland, 1
Canada, 4	Norway, 2	Romania, 1
Netherlands, 4	Palestine, 2	South Africa, 1
Sweden, 4	Australia, 1	Taiwan, 1
Austria, 3	Austrian Empire, 1	Turkey, 1
Russia, 3	Bavaria, 1	USSR, 1
Russian Empire, 3	Belgium, 1	West Germany, 1
	China, 1	

The Chemistry award is spread over 18 individual affiliation countries and twice was awarded to someone who had a professional affiliation with an institution in two different countries. By more than a factor of two, the U.S. leads the way with a whopping 87 winners. The chemistry award by birth country is a bit more diverse, covering 38 different countries of origin for the awardee. Again, the U.S. is the leader by a very wide margin. The dominance of the U.S. in the Chemistry prize (and other awards) begins shortly after World War II. Prior to that, Western Europe (as a region), was the leader. A similar observation can be made (at least with respect to the rise of the dominance by the U.S.) in other fields as well.

PHYSIOLOGY OR MEDICINE

By affiliation

U.S., 116	Denmark, 4	Netherlands, 1
U.K., 32	Canada, 3	Portugal, 1
Germany, 15	Italy, 3	Prussia, 1
France, 10	Japan, 3	Russia, 1
Switzerland, 8	Norway, 2	Spain, 1
Sweden, 7	Argentina, 1	Tunisia, 1
Australia, 4	China, 1	Germany/Japan, 1
Austria, 4	Dutch East Indies, 1	U.S. & Hungary, 1
Belgium, 4	Hungary, 1	

By birth

U.S., 81	Belgium, 3	Iceland, 1
U.K., 25	Netherlands, 3	India, 1
Germany, 20	Russian Empire, 3	Lebanon, 1
France, 13	Scotland, 3	Luxembourg, 1
Australia, 7	South Africa, 3	Mecklenburg, 1
Sweden, 7	Argentina, 2	New Zealand, 1
Switzerland, 6	China, 2	Portugal, 1
Austria, 5	Hungary, 2	Prussia, 1
Italy, 5	Norway, 2	Romania, 1
Japan, 5	Poland, 2	Russia, 1
Canada, 4	Spain, 2	Venezuela, 1
Denmark, 4	Austrian Empire, 1	Württemberg, 1
Austria-Hungary, 3	Brazil, 1	

The Physiology or Medicine award has been awarded to individuals affiliated with 24 different countries, hailing from 38 different countries. Once again, the U.S. dominates the field. As was seen with the birth countries of some of the chemistry Laureates, here we see countries that are no longer on the map. We also see how countries are referred to (e.g., Russian Empire vs. Russia and elsewhere U.S.S.R.) changing. This is deliberately done in this list to try to provide the best possible historical context of the number of awards. What countries were known as during the award gives readers insight into the timeline and geopolitical climate of the respective eras.

PHYSICS

By affiliation

U.S., 108	U.S.S.R., 7	Austria, 1
U.K., 27	Sweden, 5	Belgium, 1
Germany, 21	Canada, 3	China, 1
France, 14	Denmark, 3	Germany & U.S., 1
Switzerland, 9	Italy, 3	India, 1
Japan, 7	Russia, 2	Ireland, 1
Netherlands, 7	Australia, 1	U.S. & Japan, 1

By birth

U.S., 72	Sweden, 4	Czechoslovakia, 1
Germany, 24	West Germany, 4	French Algeria, 1
U.K., 23	India, 3	Hesse, Kassel, 1
Japan, 12	U.S.S.R., 3	Ireland, 1
France, 11	Australia, 2	Luxembourg, 1
Netherlands, 9	Austria-Hungary, 2	Morocco, 1
Canada, 6	Austria, 3	Norway, 1

PHYSICS

Italy, 6	Hungary, 3	Russian Empire, 1
Russia, 6	Denmark, 2	Schleswig, 1
Switzerland, 6	Poland, 2	Scotland, 1
China, 5	Russian Empire, 2	
Prussia, 4	Belgium, 1	

The award for physics has been awarded to individuals affiliated with 19 different countries with awards being given to someone with multiple affiliations. As was the case in both the Chemistry and Physiology or Medicine awards, the number of birth countries ticks higher, this time at 34. This is at least in part due to some of these countries ceasing to exist. Another part, however, is that people—including research scientists—relocate to other countries. In fact, after World War II, the United States "imported" (some may use the word poached) many topflight physicists from former enemy countries such as Germany.

PEACE

By affiliation/residence

Organization, 29	Canada, 2	Costa Rica, 1
U.S., 21	East Timor, 2	Democratic Republic of
U.K., 12	Egypt, 2	Congo, 1
France, 9	Iran, 2	Denmark, 1
Sweden, 5	Israel, 2	Ethiopia, 1
Germany, 4	Liberia, 2	Finland, 1
South Africa, 4	Norway, 2	Ghana, 1
Switzerland, 4	U.S.S.R., 2	Guatemala, 1
Belgium, 3	Bangladesh, 1	Iraq, 1
India, 3	Belarus, 1	Italy, 1
Ireland, 3	Burma, 1	Japan, 1
Argentina, 2	China, 1	Kenya, 1
Austria, 2	Colombia, 1	Netherlands, 1
Palestine, 1	Russia, 1	Yemen, 1
Philippines, 1	South Korea, 1	
Poland, 1	Vietnam, 1	

By birth

U.S., 19	Russian Empire, 2	Gold Coast, 1
France, 8	Scotland, 2	Guatemala, 1
U.K., 8	U.S.S.R., 2	India, 1
Germany, 6	Austria, 1	Iraq, 1
Sweden, 5	Austrian Empire, 1	Japan, 1

PEACE

South Africa, 4	Belarus, 1	Kenya, 1
Switzerland, 4	British India, 1	Netherlands, 1
Belgium, 3	Burma, 1	Ottoman Empire, 1
Egypt, 3	Canada, 1	Pakistan, 1
Ireland, 3	China, 1	Philippines, 1
Argentina, 2	Colombia, 1	Romania, 1
East Timor, 2	Costa Rica, 1	Russia, 1
Iran, 2	Denmark, 1	South Korea, 1
Liberia, 2	Ethiopia, 1	Tibet, 1
Norway, 2	Ethiopian Congo, 1	Vietnam, 1
Poland, 2	Finland, 1	Yemen, 1

Peace, the prize with the greatest diversity of all kinds is one of the two award areas where the United States does not dominate. Part of why is that the leader is not any one country but an organization. I have chosen to not try to parse out the country of origin of these organizations since it seems more important to me that the award went to an organization, rather than an individual compared to that organization's country of origin. As far as individuals go, however, the U.S. retains its spot at the top, even if by a slimmer than used to margin. The award has been given to individuals affiliated with or living in 44 different countries at the time of the award with a birth country number of 48. Of the prizes considered so far—and Literature will have this phenomenon as well—Peace is different in that the winner is not always associated with some manner of academic or research institution, or other part of research or industrial science. In fact, both Literature and Peace are first and second places in all measures of diversity. The award for Peace has gone to pacifists, civil rights activists, religious figures, and politicians/heads of state. As stated earlier, high profile wealthy institutions seem to benefit, maybe even selected for lack of diversity. In the two awards where this is most apt to being neutralized, diversity is greater. It does not reach the diversity of humanity itself, but it is measurably more diverse.

LITERATURE

By affiliation/residence

France, 18	Norway, 3	Guatemala, 1
U.K., 13	U.S.S.R., 3	Hungary, 1
U.S., 12	Chile, 2	India, 1
Switzerland, 12	Greece, 2	Israel, 1
Germany, 8	Japan, 2	Nigeria-1
Sweden, 8	Mexico, 2	Peru & Spain, 1
Italy, 6	South Africa, 2	Poland & U.S., 1
Spain, 5	Belgium, 1	Portugal, 1

LITERATURE

Poland, 4	Canada, 1	St. Lucia, 1
Austria, 3	Czechoslovakia, 1	Turkey, 1
Denmark, 3	Egypt, 1	Yugoslavia, 1
Ireland, 3	Finland, 1	

By birth

France, 12	Chile, 2	Ottoman Empire, 1
U.S., 10	China, 2	Persia, 1
Germany, 7	India, 2	Peru, 1
Sweden, 7	South Africa, 2	Portugal, 1
Spain, 6	Belgium, 1	Prussia, 1
U.K., 6	Bosnia, 1	Romania, 1
Italy, 5	Bulgaria, 1	Schleswig, 1
Denmark, 4	Crete, 1	St. Lucia, 1
Ireland, 4	Colombia, 1	Switzerland, 1
Russia, 4	Egypt, 1	Tanzania, 1
Russian Empire, 4	French Algeria, 1	Trinidad and
Japan, 3	Guadeloupe Island, 1	Tobago, 1
Norway, 3	Hungary, 1	Turkey, 1
Poland, 3	Iceland, 1	Tuscany, 1
Austria, 2	Madagascar, 1	Ukraine, 1
Austria-Hungary, 2	Mexico, 1	U.S.S.R., 1
Canada, 2	Nigeria, 1	

The prize for literature is the second award area where the U.S. does not have dominance. Here, both France and the U.K. surpass the U.S.'s count and Switzerland matches it. All told, affiliation countries ring in at 35 and birth countries at 39.

ECONOMIC SCIENCES

By affiliation

U.S., 72	Norway, 2	Netherlands, 1
U.K., 7	Sweden, 2	U.S.S.R., 1
France, 3	Denmark, 1	
Germany, 3	Finland, 1	

By birth

U.S., 60	Russia, 2	Germany, 1
France, 4	Russian Empire, 2	Hungary, 1
U.K., 4	Sweden, 2	Israel, 1
Canada, 3	Austria, 1	Palestine, 1
Netherlands, 3	British Mandate of Palestine, 1	Scotland, 1
Norway, 3	British West Indies, 1	
India, 2	Cyprus, 1	

In Economic Sciences, the dominance flexed by the U.S. borders on obscene. Greater than 70% of the awards have been won by someone affiliated with the United States. Affiliation countries ring in at a mere 10 and birth countries at 19. The smaller numbers of overall countries are no surprise since this award has existed for roughly half the time of the other awards.

Another trend is observed if one considers the prize statistics more carefully and this trend too is observed across multiple disciplines as well with exceptions being Literature and Peace. Most pronounced in Chemistry, earlier in the history of the Nobel Prize, it was less common, though not unheard of for there to be more than one awardee in any given year. The award going to more than one person is increasingly common. Whether this reflects an attempt to acknowledge more of the contributors, indecision, or a nod of sorts to the increasingly blended nature of sciences is difficult to pin down at this time but it is a curious observation, nonetheless. Another possible explanation is that there is simply more science being done and with that increase in activity, and thus there is a logical increase in the number of awards.

ALL PRIZES

By affiliation

U.S., 415	Finland, 4	Ethiopia, 1
U.K., 120	Russia, 4	Germany and Japan
Germany, 84	China, 3	Germany & U.S., 1
France, 63	Egypt, 3	Ghana, 1
Switzerland, 39	Chile, 2	Iraq, 1
Sweden, 32	Czechoslovakia, 2	Kenya, 1
Organization, 27	East Timor, 2	Nigeria, 1
Japan, 19	Greece, 2	Palestine, 1
Italy, 14	Guatemala, 2	Peru & Spain, 1
U.S.S.R., 14	Hungary, 2	Philippines, 1
Denmark, 14	Iran, 2	Poland & U.S., 1
Austria, 12	Liberia, 2	Prussia, 1
Canada, 12	Mexico, 2	South Korea, 1
Netherlands, 11	Portugal, 2	St. Lucia, 1
Belgium, 9	Bangladesh, 1	Switzerland & U.S., 1
Ireland, 7	Belarus, 1	Tunisia, 1
Israel, 7	Belgium & U.S., 1	Turkey, 1
South Africa, 6	Burma, 1	U.S. & Hungary, 1
Spain, 6	Colombia, 1	U.S. & Japan, 1
Australia, 5	Costa Rica, 1	Vietnam, 1
India, 5	Democratic Republic of Congo, 1	Yemen, 1
Poland, 5	Dutch East Indies, 1	Yugoslavia, 1
Argentina, 4		

ALL PRIZES

By birth

U.S., 299	U.K., 91	Germany, 82
France, 60	British Mandate of	Yemen, 1
Sweden, 29	Palestine, 3	Bavaria, 1
Japan, 28	Romania, 3	Korea, 1
Switzerland, 20	Chile, 2	Lebanon, 1
Canada, 20	East Timor, 2	Mecklenburg, 1
Netherlands, 20	Iran, 2	Taiwan, 1
Italy, 17	Liberia, 2	Venezuela, 1
Russia, 17	Mexico, 2	Württemberg, 1
Russian Empire, 17	Portugal, 2	Hesse-Kassel, 1
Austria, 15	Colombia, 2	Morocco, 1
Austria-Hungary, 13	Turkey, 2	British India, 1
Denmark, 13	Iceland, 2	Ethiopian Congo, 1
Norway, 13	Luxembourg, 2	Gold Coast, 1
China, 11	New Zealand, 2	Pakistan, 1
Prussia, 11	French Algeria, 2	Tibet, 1
South Africa, 10	Schleswig, 2	Bosnia, 1
Australia, 10	Ottoman Empire, 2	Bulgaria, 1
India, 10	Israel, 1	Crete
Poland, 10	Finland, 1	Guadeloupe Island, 1
Scotland, 10	Czechoslovakia, 1	Madagascar, 1
Belgium, 9	Guatemala, 1	Persia, 1
Hungary, 8	Belarus, 1	Peru, 1
Ireland, 8	Burma, 1	Tanzania, 1
Spain, 8	Costa Rica, 1	Trinidad and Tobago, 1
U.S.S.R., 7	Ethiopia, 1	Tuscany, 1
Egypt, 6	Iraq, 1	Ukraine, 1
West Germany, 5	Kenya, 1	British West Indies, 1
Argentina, 4	Nigeria, 1	Cyprus, 1
Palestine, 3	Philippines, 1	
Austrian Empire, 3	South Korea, 1	
	St. Lucia, 1	
	Vietnam, 1	

When considering a compilation of all the awards, a total of 65 countries have had someone affiliated win the Nobel Prize with birth countries totaling 92 different countries. Currently, the U.S. has a healthy and likely uncatchable lead in both the affiliation number and birth number. It is worth noting, however, that there is a disparity of just over 100 in the affiliation number vs. birth number for U.S. awardees with birth trailing. This difference amounts to just over 25% of the U.S. awardees. This means that a healthy proportion of U.S.-based winners are immigrants of some kind. The end result of this reality is

that just over ¼ of the prestige enjoyed by U.S.-based institutions are because
of immigrants.

WHAT TO DO?

As it is virtually always the case that other work is at last as worthy as the
awarded discovery, it is inevitable that this snubbed work was done by one
or more women or someone with non-western origins. Without question, part
of the problem is that some of these potentially deserving scientists are not
even nominated. Since no contemporary list of nominations is published, it is
impossible to know unless a committee member inappropriately leaks information, or a nominator discloses their nomination. The online nomination
archive mentioned earlier is a wonderful step toward some measure of (even
if only historical) transparency. Although causes for this potential and in some
cases actual lack of nomination are likely to be many, it is nearly impossible
to exclude sexism and/or racism as factors. While we in the sciences may like
to believe that we are above prejudices like racism, sexism, and all the other
"isms" in our society, focusing our attention instead on independent interpretations of data, we in fact are not immune. Personal conflicts also inevitably
rear their ugly heads. There are multiple pathways that may be effective at
correcting this. For example, opening the nomination process more broadly
or at least disclosing nominees may be low-hanging fruit for a more equitable
process, assuming that the cause is a lack of nominations for otherwise worthy
awardees. This is because it would bring about at least a small measure of putative accountability through better transparency.

  As an academic, I like rubrics, a lot. A well-designed rubric that generates
nominations for the committee may better recognize the work of these snubbed
scientists. Even the award process could be made more fair, consistent, and
equitable using a rubric. The internet and data analysis allow for all sorts of
metrics to be accessed and assessed. Metrics include the number of publications;
the number of citations, including citations per publication; and so many
others can be found by anyone literate with an internet connection; much of
the data isn't even behind a paywall. Such an approach would be defensible,
consistent, and blind to every attribute of the researcher except for the work
itself. Although it is possible that such a rigid evaluation method is already
practiced, the small number of women awardees casts doubt, serious doubt in
my opinion, on this possibility. Another flaw that can be argued is that work
that is infamous or debunked may at times get enough "attention" by way of
negative citations, that the rubric identifies it as worthy. This is where a human

element would be able to (and ought to) override the rubrics. That said, opposition to such a rigid approach is not unfair. It can certainly be argued that such a rubric will disqualify a "dark horse" awardee. I at once disagree with this and find it irrelevant even if it turns out to be right. First, a well-designed rubric can easily allow for such a nomination and award. Second, the Nobel Prize is not the NCAA Basketball Tournament, March Madness™. Nor is it the playoffs for some professional sport or even the Olympics where "Cinderella Stories" enthrall (and disappoint) millions. Although sometimes it takes years, decades even, to fully recognize and appreciate the enormous impact and importance, the Nobel Prize is not the place for a *feel-good underdog story*. It is for this reason that I believe a rubric that in some way evaluates the *already realized* impact of the works being nominated and considered is the fairest way forward. Quotas are one option that has been dismissed,[21] a dismissal I agree with.

FUTURE OF THE NOBEL PRIZE

There are no indications the Nobel Prize will cease to exist any time soon and I cannot help but worry that the Peace Prize, especially, will slowly but surely be increasingly accused of being political, but this is unlikely to present a challenge to its existence. Increasingly, some of the science-based awards are taking on more of a lifetime achievement award feel than one awarded for breakthroughs made in the preceding year; this does not take away from the prize.

Increasingly controversial is who receives recognition for the work contributing to the Nobel Prize and it is here I think some changes may come eventually. This issue may even be a "threat" to the Nobel Prize as we know it. Although the Literature prize and perhaps the Peace prize are justifiably a truly individual award, scientific endeavors are more collaborative today than in Nobel's time. I would argue, in fact, that Nobel and his contemporaries could not have foreseen just how collaborative science would become. Not only are projects often conducted by teams of several researchers, for multiple years rather than an individual, the nature of any individual project is also increasingly collaborative since modern science is increasingly inter/multidisciplinary. As a result, it often demands multiple fields of expertise, too many fields for any *one* person. This inevitably leads to a larger number of people. Projects also now take far longer to complete than 120 years ago—at least in part because of higher publication standards—and the increase in time only further increases the number of people involved. This brings about a natural, and exceedingly difficult to answer question of who among all these collaborators deserve(s) the award. Often, the PI—the boss for the reader unfamiliar with the terminology—is the one who receives the award. This is true, even though in modern science this individual is very unlikely to have performed even one of the experiments behind the work. Why then do they get the award? It is as complicated as it sounds unfair.

First and foremost, this individual is *the constant*; they are the person who has been involved in the project from the beginning and is the only person (usually, anyway) who has had input to add to every aspect of the project since its inception. More for the inexperienced reader than the experienced reader, this individual is also the person who pays most of the (scientific) bills for the research through their grants. They also, whether they are in the lab doing actual scientific work or not, contribute heavily to the projects by way of suggestions for future/additional experiments and problem-solving. That said, there is some validity to complaints that such an approach is unfair and that it fails to appropriately acknowledge the work of the people performing the experiments in the laboratory. Unfortunately, I do not see any viable alternatives currently, though it is for sure something to attempt to right. There has even been a suggestion that the Nobel Prize should be given for a topic, rather than to people.[22] How this relates to decidedly individual prizes such as Peace and especially Literature is unclear to me. It would create an altogether different disparity if some Nobels are treated one way and the others an alternative way. The goal of fairer recognition is good, noble even, but to create a disparity to do so would be wrongheaded, I think. I at the same time concede that a good reason to make different awards consistently executed is lacking. That said, to make changes necessary to see true progress in righting this would violate the terms in Nobel's will, admittedly now over a century old.

The Chemistry Prize

Since its inception, through April 2024, the Chemistry award has been made 116 times. In 1916, 1917, 1919, 1924, 1933, 1940, 1941, and 1942, no award was given. Two people, Frederik Sanger, and K. Barry Sharpless have been awarded the Chemistry prize twice, though Linus Pauling and Marie Curie are chemistry Laureates who have won two awards in a different field. Both of Pauling's awards (Chemistry and Peace) were unshared and he remains the only *individual* to win more than one *unshared* Nobel Prize of any kind. Curie, meanwhile, is one of two women to have earned an unshared chemistry award, the other being Dorothy Crowfoot Hodgkin in 1964 for her work on solving the structure of important biological substances using X-rays. Meanwhile, 60 men have taken home the prize solo. Once again, it is hard to justify this sort of disparity, even keeping in mind the growing number of instances where the award is given to more than one person. Curie remains the only woman to win more than one Nobel; one for physics (1903) and one for chemistry (1911).

The chemistry medal is the handiwork of Swedish sculptor and engraver Erik Lindberg. It represents Nature in the form of a goddess resembling Isis. In her arms, she holds a cornucopia and is emerging from the clouds, a veil, covering her face, is held up by the Genius of Science.

Future of the Chemistry Nobel Prize

A survey of what, specifically the Nobel Prize in chemistry, is awarded for shows a clear rise of biochemistry (Table 1.4). Anecdotally, a colleague of mine has lamented that the Chemistry Nobel Prize seems to them to be more biology sometimes than chemistry. The Physiology or Medicine prize is also often as much biochemistry as it is physiology or medicine. Time will tell if this trend continues or shifts to more energy-based science.

Predictions

If for no other reason but to have a little fun, I would like to make a few predictions on future Nobel Prizes. For soothsaying, I am going to focus my predictions to the Physics, Chemistry, and Physiology or Medicine awards, though I will comment on the Peace and Literature prizes briefly as well. Regarding the Chemistry Nobel Prize, as the synthetic protocols, currently referred to as C–H activation, see further development, it is highly likely to earn someone a Nobel Prize. It is currently too early to identify a true front-runner, but Melanie Sanford and M. Christina White are currently two leaders in the field. The former has also been active in fluorination reactions and flow batteries. Fluorinated compounds are important in pharmaceutical compounds (as therapeutics, and positron emission tomography (PET) imaging agents) and agrochemicals. Meanwhile, flow batteries hold promise as grid-scale storage devices for renewable energy generation and storage.

Martin Burke has been active in—among other areas—generating methods for Suzuki coupling reactions that could revolutionize how C–C bonds are made[23] for complex syntheses. However, the use of palladium-catalyzed reactions is problematic for the large scales needed in ton-scale synthesis of pharmaceutical compounds, a reality that may dampen the broad applicability of such protocols.

Burke has also been active in attempting to use machine learning to optimize reaction conditions.[24] Although this study failed to identify a standard protocol, with the rapid expansion of AI likely to come, attempts to use machine learning and AI to generate standard protocols have likely only begun. Burke and his lab speculate this particular work failed because synthetic protocols are likely to be borne out of subjective preferences by the authors who reported the protocols or other biasing decision-making influencers like availability of solvents and other reagents and/or negative data. They claim, rightly I suspect, that to be successful, algorithms will need training with very carefully selected data sets.

Alternatively, carbon sequestration or the conversion of carbon dioxide back to gasoline or other oil products are also important endeavors that would be the carbon and environmental equivalent to Haber's Nobel-winning work. Although it is without question worth asking, "Why would we remake oil products from carbon dioxide?," it is important to recognize that the oil industry provides way more than just energy. It also provides starting materials and solvents for various critical materials such as textiles, dyes, and arguably most important of all, the pharmaceutical industry. An alternative source of these important substances or of fuel will be very important as these non-renewable resources dwindle in their supply and/or access due to geopolitical conflicts and increase in their cost.

Daniel Nocera has made great strides in this area with the artificial leaf and its successor, the bionic leaf. In short, these products purport to produce food and fuel using only sunlight, air, and water. Essentially between the two, these products sequester carbon dioxide, fix nitrogen to ammonia, and generate hydrogen gas and oxygen gas using *any* water source, air, and sunlight. The implications go far beyond just energy generation, which the hydrogen certainly does, but also can replace commercial fertilizer, the generation of which consumes tremendous amounts of energy and generates a considerable amount of carbon dioxide worldwide. Nocera's products could replace commercial fertilizer in at least some cases if all proofs of concept come to fruition, all the while removing carbon dioxide from the atmosphere. To say that these have the potential to change, and I dare say save the world is an understatement.

Taking physics next, if life—even simple life—were ever to be discovered elsewhere in the solar system (e.g., Mars or some moon), those responsible for doing the experiments or at least those who designed them will almost certainly win a Nobel Prize. Such a discovery would have incomprehensible ramifications for humankind. If recreational space travel becomes more popular, it is plausible (in my opinion, anyway) that either Elon Musk (*via* Space X) or Jeff Bezos (*via* Blue Origin) may be in line for a Physics Nobel Prize. In fact, what those two have done already may be worthy of a Nobel Prize. The development of the technologies by an independent company rather than governmental entity such as NASA or any of its counterparts elsewhere is utterly amazing. The development of the technologies each company uses are as sophisticated or more as anything any Nobel Prize has been previously awarded for. The automated and reusable rockets are technological wonders to be sure. The other thing they have potentially done is popularize something related to science. As of this writing, recreational space travel accessible to the masses is a long way off. However, the first steps have been taken because of these efforts. All that said, going to the Moon did not earn anyone a Nobel Prize so it is perhaps foolhardy to think this any of will.

Other technology related to renewable energy also may eventually earn someone a Nobel Prize in physics. However, improvements that would likely be needed to make that a reality would need to be significant, either in the reduction of cost, increase of yield/conversion, or both. Actually, depending on the precise nature of the technology, this sort of breakthrough may in fact be more worthy of a Chemistry Nobel Prize than a Physics one as it would be with Nocera's. It is shocking to me that energy storage has only been part of one Nobel Prize to date, the prize in 2019 for lithium-ion batteries, awarded to John B. Goodenough, M. Stanley Whittingham, and Akira Yoshino.

The electric or self-driving car both represent technological breakthroughs absolutely worthy of a Nobel Prize and Elon Musk also has a hand in both. Each have enormous advantages and of course disadvantages over current technologies and as they improve, we are increasingly likely to reap all those benefits and more. However, lacking a Nobel Prize in computer science, it is hard to place this into one of the fields. Similar things can be said about artificial intelligence platforms such as Chat GPT.

For Physiology or Medicine, it is low-hanging fruit to predict that the m-RNA technology that led to the most effective of the COVID-19 vaccines will earn someone the Nobel Prize and in 2023, exactly this happened. Additional awards may yet come if it is found that this technology is universally applicable to other vaccines as well and I predict it will be. This would be especially true if applicable to a disease such as HIV or cancer. Extremely and/or broadly effective cancer vaccines and even man-made anticancer viruses have been developed and if they prove successful would without question be at or near the front of the line for the award.

Although well outside my professional comfort zone, Taylor Swift may eventually join Bob Dylan as a Nobel Prize in Literature winning singer-songwriter. My claim is based on the fact that Swift writes nearly all her songs and that the impact they've had on her fans is profound. In particular, her most recent (2023–2024) Eras Tour combines songwriting, performance, and visual arts almost like no other. The strongest arguments in her favor are her authorship (like Dylan) and how millions upon millions of people across cultures identify with her songs and words. More than anything else, this speaks to the power, reach, and quality of her work.

Finally, although not exactly a prediction, I must admit I am shocked that the Nobel Peace Prize has not yet been awarded to the Bill and Melinda Gates

TABLE 1.4 Summary of subdisciplines of chemistry receiving the Nobel Prize

SUBDISCIPLINE KEY: A, ANALYTICAL CHEMISTRY/INSTRUMENTATION; B, BIOCHEMISTRY; EG, ENERGY STORAGE; EV, ENVIRONMENTAL; G, GENERAL CHEMISTRY; I, INORGANIC CHEMISTRY; M, MATERIALS; N, NUCLEAR; O, ORGANIC CHEMISTRY; P, PHYSICAL CHEMISTRY

YEAR	SUBDISCIPLINE	YEAR	SUBDISCIPLINE	YEAR	SUBDISCIPLINE	YEAR	SUBDISCIPLINE	YEAR	SUBDISCIPLINE
1901	P	1902	O	1903	G	1904	G	1905	O
1906	G	1907	B	1908	N	1909	G	1910	O
1911	N	1912	O, O	1913	I	1914	G	1915	B
1916	None	1917	None	1918	I	1919	None	1920	P
1921	N	1922	A	1923	A	1924	None	1925	G
1926	G	1927	B	1928	B	1929	B, B	1930	B
1931	G, G	1932	I	1933	None	1934	N	1935	N, N
1936	A	1937	B, B	1938	B	1939	B, O	1940	None
1941	None	1942	None	1943	P	1944	N	1945	B
1946	B, B, B	1947	O	1948	A	1949	P	1950	O, O
1951	N, N	1952	A, A	1953	I	1954	G	1955	B
1956	P, P	1957	B	1958	B	1959	A	1960	EV
1961	EV	1962	B, B	1963	M, M	1964	A	1965	O
1966	G	1967	A, A, A	1968	P	1969	G, G	1970	B
1971	G	1972	B, B, B	1973	O, O	1974	P	1975	O, O
1976	I	1977	P	1978	B	1979	O, O	1980	B, B, B
1981	O, O	1982	A	1983	I	1984	O	1985	A, A
1986	P, P, P	1987	O, O, O	1988	P, P, P	1989	B, B	1990	O

(continued)

TABLE 1.4 (Continued)

SUBDISCIPLINE KEY: A, ANALYTICAL CHEMISTRY/INSTRUMENTATION; B, BIOCHEMISTRY; EG, ENERGY STORAGE; EV, ENVIRONMENTAL; G, GENERAL CHEMISTRY; I, INORGANIC CHEMISTRY; M, MATERIALS; N, NUCLEAR; O, ORGANIC CHEMISTRY; P, PHYSICAL CHEMISTRY

YEAR	SUBDISCIPLINE	YEAR	SUBDISCIPLINE	YEAR	SUBDISCIPLINE	YEAR	SUBDISCIPLINE	YEAR	SUBDISCIPLINE
1991	A	1992	P	1993	B, B	1994	O	1995	EV, EV, EV
1996	O, O, O	1997	B, B, B	1998	P, P, P	1999	A	2000	M, M, M
2001	O, O, O	2002	A, A, A	2003	B, B	2004	B, B	2005	O, O, O
2006	B	2007	M	2008	B, B, B	2009	B, B, B	2010	O, O, O
2011	I	2012	B, B	2013	G, G, G	2014	A, A, A	2015	B, B, B
2016	G, G, G	2017	B, B, B	2018	B, B, B	2019	EG, EG, EG	2020	B, B
2021	O, O	2022	O, O, O	2023	P, P, P				

foundation. The philanthropy done by this organization to date is tremendous and world changing. Because of this generous work, hundreds of millions of people in some of the poorest and most disease-ridden parts of the world have access to medicine. The foundation has also done amazing work to improve education and access to technology in classrooms, globally. My only guess as to why no Nobel Prize has been awarded to the foundation is an attempt to send a message that you cannot "buy" a Nobel Peace Prize. Nothing other than that sort of message can explain to me why this foundation has not been bestowed this (in my opinion deserving) honor.

The Prize and Society

Shortly after the announcement of the 2021 Physics Prize, an opinion appeared on CNN claiming, "This Nobel Prize is a game-changer."[25] This reflects the opinion of the article's author that—since the Nobel Prize here is related to climate change—this should turn the tides of belief to in favor of the validity of anthropogenic (human-caused) climate change. This, because of an implied vouching (my words) done by the committee for science by making this award. In my opinion, this is untrue. Those denying anthropogenic climate change are already willfully ignoring the widely agreed upon science. Though it is easy to make claims that other widely held beliefs have oft been proved wrong, the case for anthropogenic climate change is growing every year, rather than waning. The deniers seem to me to claim that "it is all a natural cycle anyway." This, to my understanding is true but it is also my understanding that all the best scientific evidence says that our activities on Earth are exacerbating or hastening (or both) these natural cycles. To date, most of the corrections science has made to its understanding of this issue is that the impacts may have been underestimated. One more group of scientists "endorsing" it is very unlikely to perturb the stance of deniers. Additionally, there are many people who (apparently without jest) believe the world is flat and/or that the world is a mere several thousand years old. Both beliefs are held even when confronted with mountains (pun very much intended) of scientific evidence. In fact, I may argue the exact opposite stance about the societal impacts of the award. The deniers may in fact lose even more faith in the scientific establishment, viewing the award as something agenda-driven, rather than science-driven. Such possibilities—turning the tides of belief, further entrenching against, or anything in between—should *never* be the goal of the award. The goal of the Nobel Prize never was, is not, and never will be to convince the throes of non-science persons to believe a scientific finding. The same should be said about the Peace Prize. It should not be awarded for the goal of calling attention to a cause. Rather, it should be awarded for work someone, or a group has done for a cause related to peace or the otherwise benefit of humankind.

THE IG®NOBEL PRIZES[26]

The website improbable.com has awarded a set of mock Nobel Prizes called the Ig®Nobel Prize since 1991. The prizes tout themselves as being awarded "for achievements that first make people laugh then make them think." I personally enjoy these a great deal and look forward to their announcement every bit as much as I look forward to the announcement of the actual Nobel Prizes. I make a point every year to share the prize announcement (for both awards) with my students. In the case of the Ig®Nobel Prizes, I do it to not only share entertainment but also encourage them to be limitless with what they consider research ideas they will pursue in their careers or in classes we share. The topics for which the Ig®Nobel prizes are awarded are inconsistent from year to year. Some of my favorites are listed here in chronological order. In all cases possible, I have provided the citation to an actual peer-reviewed paper where the project was originally described/reported.

1993, Medicine—James F. Nolan, Thomas J. Stillwell, and John P. Sands, Jr. for their *Journal of Emergency Medicine*, 1990, 305–307, paper titled "Acute Management of the Zipper-Entrapped Penis."

1995, Public Health—Martha Kold Bakkevig for her paper in *Ergonomics*, 1994, 1375–1389, titled "Impact of Wet Underwear on Thermoregulatory Responses and Thermal Comfort in the Cold."

1995, Physics—D. M. R. Georget, R. Parker, and A. C. Smith for their paper in *Powder Technology*, 1994, 189–196, titled "A Study of the Effects of Water Content on the Compaction Behaviour of Breakfast Cereal Flakes."

1996, Public Health—Ellen Kleist and Harold Moi for their paper in *Genitourinary Medicine*, 1993, 322, titled "Transmission of Gonorrhea Through an Inflatable Doll."

1996, Physics—Robert Matthews for his paper in the *European Journal of Physics* 1995, 172–176, titled "Tumblin Toast, Murphy's Law and the Fundamental Constants"

2000, Medicine—Willibrord Weijmar Schultz, Pek van Andel, and Eduard Mooyaart for their paper in *British Medical Journal*, 1999, 1596–1600 titled, "Magnetic Resonance Imaging of Male and Female Genitals During Coitus and Female Sexual Arousal."

2000, Psychology—David Dunning and Justin Kruger for their paper in *Personality and Social Psychology*, 1999, 1121–1134, titled "Unskilled and

Unaware of It" How Difficulties in Recognizing One's Own Incompetence Lead to Inflated Self-Assessments." *Note: This is now very much accepted psychology and is often talked about in education circles.*

2001–Chittaranjan Andrade and B. S. Srihari for their paper in the *Journal of Clinical Psychiatry* 2001, 426–431, titled "A Preliminary Survey of Rhinotillexomania in an Adolescent Sample." P.S. rhinotillexomania is nose picking.

2004, Public Health—Jillian Clarke-for her publication in *ACES College News* on September 2, 2003, titled "If You Drop It Should You Eat it? Scientists Weigh In on the 5-Second Rule."

2006, Physics—Basile Audoly and Sebastien Neukirch for their *Physical Review Letters* 2005, 095505, paper titled "Fragmetnation of Rods by Cascading Cracks: Why Spaghetti Does Not Break in Half."

2008, Economics—Geoffrey Miller, Joshua Tybur, and Brent Jordan for their paper in *Evolution and Human Behavior*, 2007, 375–381, titled "Ovulatory Cycle Effects on Tip Earnings by Lap Dancers: Economic Evidence for Human Estrus."

2010, Biology—Libiao Zhang, Min Tn, Guangjian Zhu, Jianping Ye, Tiyu Hong, Shanyi Zhou, Shuyi Zang, and Gareth Jones for their *PLoS ONE* vol. 4, e7595, paper titled "Fellatio by Fruit Bats Prolongs Copulation Time."

2012, Fluid Dynamics–Rouslan Krechetnikov and Hans Mayer for their *Physical Review E*, 2012, paper titled "Walking with Coffee: Why Does It Spill?."

2017, Anatomy—James Heathcote for his *British Medical Journal*, 1995, 1668, paper titled "Why Do Old Men Have Big Ears?"

2018, Chemistry—Paula Romão, Adília Alarcão, and César Viana for their *Studies in Conservation*, 1990, 153–155, paper titled "Human Saliva as a Cleaning Agent for Dirty Surfaces."

2020, Materials Science —Metin Eren, Michelle Bebber, James Norris, Alyssa Perrone, Ashley Rutkoski, Michael Wilson, and Mary Ann Raghanti for their *Journal of Arcaeological Sciences: Reports*, 2019, 102002, paper titled "Experimental Replication Shows Knives Manufactured from Frozen Human Feces Do Not Work." (Yes, you read that correctly!)

2023, Medicine—Christine Pham, Bobak Hedayati, Kiana Hashemi, Ella Csuka, Tiana Mamaghani, Margit Juhasz, Jamie Wikenheiser, and Natasha Mesinkovska for their *International Journal of Dermatology* 2002, e456–457, paper titled "The Quantification and Measurement of Nasal Hairs in a Cadaveric Population."

You may ask why I share such borderline nonsense here and in my classes. That's a fair question. I do so for more than one reason but the two most important ones are that I believe strongly in enjoying ourselves in classes and in what we read. These studies give science some measure of personality rather than continue a caricature of *humorless nerds*; it shows us for the *comical nerds* some of us are. Finally and more importantly, I do so to encourage students to be curious and explore their ideas *no matter how crazy* they think their ideas are.

ABOUT THIS SERIES OF BOOKS

This series of books focuses on the Nobel Prizes in Chemistry that have contributed to the field of synthetic organic chemistry. Such a broad scope necessitates decisions that may appear arbitrary to even the learned reader. I assure you, no slight is meant by any decisions to exclude certain awards, especially those in the next section. For certain, awards other than those covered in these volumes *involve* organic chemistry, even synthetic organic chemistry. Herein, I have chosen to omit any that only utilize or apply synthetic organic chemistry, rather than build it up. A section covering "honorable mention" of Nobel recipients attempts to justify some of the omissions. In essence, I have tried to create a story of how the Nobel Prize in chemistry made synthetic organic chemistry what it is today. Are such lines blurry? In a word, yes, and to be fair, even I think some of the awards I have chosen to include are borderline, at best. As the field continues to develop, I yield that some currently omitted studies may merit inclusion. If such a scenario comes about, I will address it with later editions, volumes, and/or both.

Currently, the volumes are delineated chronologically. Such organization was chosen to allow for the easier creation of both later, updated editions and future volumes. Furthermore, a thematic organization would be difficult, if not impossible, to balance and adhere to parameters of logical and comparable volume sizes. Additionally, some years where the prize is awarded for multiple works, one may be more related to organic synthesis than the other. In this sort of case, unrelated work will be mentioned but not receive a detailed review. The mention will be restricted to the bare minimum necessary to retain historical accuracy for the award.

Volume 1
   1902—Fischer, "in recognition of the extraordinary services he has rendered by his work on sugar and purine syntheses"
   1910—Wallach, "in recognition of his services to organic chemistry and the chemical industry by his pioneer work in the field of alicyclic compounds"

1912—Grignard, "for the discovery of the so-called Grignard reagent, which in recent years has greatly advanced the progress of organic chemistry"

and Sabatier, "for his method of hydrogenating organic compounds in the presence of finely disintegrated metals whereby the progress of organic chemistry has been greatly advanced in recent years"

1950—Diels and Alder, "for their discovery and development of the diene synthesis"

1965—Woodward, "for his outstanding achievements in the art of organic synthesis"

Volume 2

1979—Brown, "for the development and of the use of boron-containing compounds into important reagents in organic synthesis"

and Wittig, "for the development and of the use of phosphorus-containing compounds respectively into important reagents in organic synthesis"

1981—Fukui and Hoffmann, "for their theories, developed independently, concerning the course of chemical reactions"

1990—Corey, "for his development of the theory and methodology of organic synthesis"

Volume 3

2001—Knowles and Noyori, "for their development of catalytic asymmetric synthesis" and "for their work on chirally catalyzed hydrogenation reactions"

and Sharpless, "for his work on chirally catalyzed oxidation reactions"

2005—Chauvin, Grubbs, and Schrock, "for development of the metathesis method in organic synthesis"

2010—Heck, Negishi, Suzuki, "for palladium-catalyzed cross couplings in organic synthesis"

Volume 4

2018—Frances H. Arnold, "for the directed evolution of enzymes" (shared with **Smith and Winter** "for the phage display of peptides and antibodies," both of whom are not covered here)

2021—List and **MacMillan**, "for the development of asymmetric organocatalysis"

2022—Bertozzi, Sharpless, and Meldal, "for the development of click chemistry and biorthogonal chemistry"

HONORABLE MENTIONS

As one might imagine, choosing who to include and more importantly *not* include in this sort of compilation is a challenging task. I dare (jokingly) quip that this decision is even more difficult than awarding a Nobel in the first place as I am assuming the position of omitting recognized achievement. Perhaps only to relieve a guilty conscience of sorts, I mention—and attempt to justify—here a handful or so of my most difficult decisions to omit individuals in chronological order.

1905—Adolf von Baeyer, "in recognition of his services to the advancement of organic chemistry and the chemical industry, through his work on organic dyes and hydroaromatic compounds." Arguably, this omission is the most egregious. His contributions to the field of organic chemistry are without doubt great and far-reaching. However, as far as their improving or building up the field of synthetic organic chemistry, not only is his primary contribution to this field (the Baeyer–Villiger reaction) not related to his Nobel, but his award-winning work does also not offer new and versatile synthetic methods or a changing of the way synthetic chemistry is conceived and performed. Perhaps future editions will include Baeyer, for now, I stand by my decision to omit.

1923—Fritz Pregl, "for his invention of the method of microanalysis of organic substances" (for chemical composition). The importance of microanalysis of organic substances cannot be overstated. This work allowed for nothing less than the far easier determination of the molecular formula of chemical substances. Although running the experiment may not be, performing the calculations using data from this experiment is a standard part of many, if not all, college-level general chemistry classes. Without such data (the molecular formula), identifying the chemical structure of compounds would be far more arduous. In fact, I don't know how it could be reliably done without the molecular formula, even if newer and better methods exist to determine it today. All that taken into consideration, this does not contribute to synthetic organic chemistry, so it is omitted here.

1947—Sir Robert Robinson, "for his investigations on plant products of biological importance, especially the alkaloids." This work showed troponin alkaloids can be made from three simpler molecules. Omitting Robinson *may* even be more egregious than omitting Bayer. What Robison (effectively) showed is complicated molecules (at least the troponin alkaloids) could be made from much simpler building blocks. Had R. B. Woodward and later Corey not *totally* changed and expanded what was possible, it would be harder to omit Robinson. The outstanding work by Woodward and Corey, however, sets the bar so much higher that it is too difficult to include Robinson.

1956—Hinshelwood and Semenov, "for their researches into the mechanism of chemical reactions." It is inarguable that a better understanding of chemical reaction mechanisms—how on an atomic level and/or molecular level molecules rearrange to go from starting materials to products for the less experienced reader—helped to drive the field of organic synthesis forward. However, as it is difficult or impossible to identify a specific way that this work built synthetic organic chemistry, it is omitted here.

1984—Robert Bruce Merrifield, "for his development of methodology for chemical synthesis on a solid matrix." Merrifield's award is basically for polymer-bound peptide synthesis and without question has totally changed peptide synthesis. Indeed, his contribution has revolutionized synthesis. However, to date, this has been primarily focused on peptide synthesis quite narrowly. If it were to ever be expanded to more general synthetic organic methods, it would be harder to justify omission. Some work has been done to apply it to general organic synthesis but in my opinion, more must be done for inclusion in this collection.

2018—Smith and Winter, "for the phage display of peptides and antibodies" while including Arnold "for the directed evolution of enzymes." In short, Arnold's work has potential to lead to enzymes being developed that will permit easier chemical transformations, perhaps—in a best-case scenario—leading to reactions that can compete with the selectivities and high yields observed with biological systems *in-vivo*. Meanwhile, the work of Smith and Winter has no relation to synthetic chemistry.

VOLUME 3 WINNER BIOGRAPHIES

This volume is different from the first two, or rather, the prizes covered herein are different. In this volume, all three awards were won by three people and all persons were awarded for contributions that build up synthetic organic chemistry. As a result, there is significantly more science to cover. These bios are done in an attempt to better humanize the winners and I hope that future generations of scientists take inspiration from not just the chemical but the personal stories as well of these giants of chemistry.

William Knowles

Knowles was born in 1917 and raised in Massachusetts, U.S. He went to a boarding school for High School and even took a second senior year before starting his undergraduate studies at Harvard. It was in boarding school that he first encountered chemistry intimately. At Harvard, Knowles muses that he earned solid Bs, rather than the straight As of many of his classmates. This didn't seem to hold him back since after completing his undergraduate studies at Harvard in 1939, he went on to Columbia to earn his PhD under Bob Elderifield (and of course later win a Nobel Prize). During his studies, Knowles suffered a setback when an intermediate he had worked on for months was destroyed in a diazomethane explosion during his time at Columbia. He is not the first, nor will he be the last I am sure, to have an accident set back studies. Unlike others, including another Nobel Prize winner, Knowles was not injured in this accident. The draft board compelled Columbia to award him his PhD in 1942 (a bit early) and he immediately took up work at the Thomas and Hochwalt Laboratories that had recently joined Monsanto. He was put to work supporting the war effort by making super pure precursor to the explosive cyclonite. Knowles also worked with Woodward on the cortisone synthesis while at Monsanto before turning his attention to kinetic studies and stayed at Monsanto until he retired in 1981. He passed away in 2012.

Ryōji Noyori

Ryōji Noyori was born in Japan in 1938 near Kobe, where he eventually lived. In his youth, he enjoyed many outdoor activities. Noyori found inspiration from the poverty of postwar Japan and had the dream to become a leading chemist to contribute to society by inventing beneficial products after attending a public conference about nylon. In middle and high school, his appetite for chemistry was furthered by Dr. Kazuo Nakamoto, from whom he received his first official chemistry lesson. In his free time, he even studied judo. In 1957, he started at Kyoto University, eventually earning his BS in 1961 and his masters in 1963. He then took a position as instructor of Professor Hitosi Nozaki's labs at Kyoto and in 1967 earned his DEng. Next, he was invited in 1967 to chair a newly created organic chemistry lab at Nogoya University, causing him to abort a plan to work as a postdoc in E. J. Corey's lab. Because of his age and inexperience, he was initially appointed as an Associate Professor. He then got into organometallics to try to create a research track independent from a senior colleague's research. In 1964 he moved to Harvard for a year learning a lot from Corey and his lab and then went back to Nagoya where he remained, working on homogenous catalysis via organometallics.

Barry Sharpless

K. (Karl) Barry Sharpless was born in 1941 in the U.S. and grew up in the Philadelphia, Pennsylvania, U.S. area. In his youth, he loved fishing and spent much of his time on the Manasquan River, often running eel and crab traps too. Sharpless found work at a young age, working on a family member's boat, something he joked his family member did to avoid hiring a full-grown person. In 1963, Sharpless completed his undergraduate studies at Dartmouth College. At the end of these studies his undergraduate mentor, Tom Spencer, convinced him to try graduate school before going on to medical school. In Sharpless's words, if he looked back at all, it was to fishing. After completing his PhD under Eugene E. von Tamelin at Stanford in 1968, he went on to two postdocs, one in 1968 and another in 1969; the first with James Collman in the area of inorganic and organometallics at Stanford and then in enzymology with Konrad Bloch at Harvard. His independent career then began at MIT from 1970 to 1977, moving to Stanford for three years before returning to MIT from 1980 to 1990. In 1990, he joined the Scripps Research Institute in La Jolla, California, where he remains today as a member of the Skaggs Institute for Chemical Biology at Scrips, a position he has held since 1996. Sharpless won his first Nobel Prize in 2001 and his second in 2023 (see Volume 4 of this series). Unlike Knowles, who escaped injury when an explosion occurred, Sharpless was not so lucky. He suffered an eye injury so severe when a tube burst that he now has a glass eye.

Yves Chauvin

Yves Chauvin was born in Belgium on October 10, 1930 and died on January 27, 2015 in France. One of Chauvin's grandmothers studied piano under Emmanual Chabier. His father went to war twice and was even taken prisoner in 1939. He credits surviving during the difficulties of war as the reason he was never a fussy eater. He pursued chemistry but his own eventual military service in part prevented him from getting his PhD. In his first job in industry, he found that process development (at the time anyway) consisted mainly of copying what already existed. Chauvin wanted something more interesting and new so he resigned. He then took a position at the Institut Françias du Pétrole, finding work he considered interesting. It was here that he began studies in the field of coordination chemistry, organometallics, and homogenous transition metal catalysis. It was through this work that he developed his first two homogeneous catalytic processes, one of which dimerized ethylene to 1-butene. It is this case that is most directly related to the work of Grubbs and Schrock who he shared the Nobel Prize with.

Robert H. Grubbs

Robert Grubbs was born on February, 27, 1942 in the U.S. and died on December 19, 2021 in the U.S. Grubbs grew up in a house his father had built mostly himself on land given as a wedding present by his father (Robert's grandfather). Grubbs grew up surrounded by a support network of uncles, aunts, and cousins. A well-educated grandmother set a high intellectual standard for all her grandchildren, leading to many becoming educators of some sort. In his youth, young Robert helped his father rebuild car engines, install plumbing, build houses, and work on his uncle's farms. Grubbs parlayed some of this experience into early jobs that helped finance his college studies. As often seems to be the case, Grubbs points to a teacher who sparked his interest in science—Mrs. Baumgardner, a junior high school teacher. Grubbs initially planned to pursue agricultural chemistry at the University of Florida in chasing this interest. He had even taken up a job working in an animal nutrition lab analyzing animal feces. One night, he joined a friend working in the organic lab of then new faculty member Merle Battiste. Grubbs found organic chemistry to smell way better (which is saying something) and never looked back. After finishing his undergraduate degree, he went on to Columbia to get his PhD with Ronald Breslow, Battiste's own PhD mentor. Around that time, Grubbs heard a talk by Rolli Pettit on the stabilization of unstable molecules by coordination to transition metals, a talk Grubbs credits with getting him interested in the topic. He then moved on to an NIH Fellowship with Jim Collman at Stanford and in 1969 started his independent career at Michigan State before moving to Caltech in 1978 where he remained until his death.

Richard R. Schrock

Richard Schrock was born in 1954 in the U.S., coming from a large family (his father was one of six and his mother was one of ten). His father, a carpenter, did many renovations on one of their family homes and the family had a prolific garden on their acre of land. Richard garnered from his father an interest in woodworking. Unlike many others in this volume, rather than a teacher, Schrock started a love for chemistry when his older brother gifted him a chemistry set at eight years old. Once 13 himself, the younger Schrock took chemistry with Harry Dailey in High School and his interest was only further stoked. Schrock even went so far as to have a small "lab" in his basement. Upon moving to California prior to finishing high school, he expanded his "lab" and carried out processes such as generating bromine from potassium bromide and sulfuric acid and chlorine gas from electrolyzing molten sodium chloride (table salt). He also did more "typical" west coast activities such as skin diving and surfing. He even put his woodworking skills to use making surfboard fins that he sold and eventually completed his undergraduate studies at the University of California at Riverside. He then went on to earn his PhD at Harvard under John Osborn and after Harvard, completed an NSF Fellowship at Cambridge with Jack (later Lord) Lewis. Upon returning to the U.S. in 1972 he took work at DuPont where he worked with Fred Tebbe. It was here that he first heard of olefin metathesis during a talk, and he became interested in it, seeing potential relations to the alkylidene complexes he had previously discovered. After inviting Sharpless to DuPont for a talk, an offer to move to MIT came and he began work there, where he remains.

Richard Heck

Richard Heck was born August 15, 1931 in the U.S. and died on October 10, 2015 in the Philippines. At the age of eight, Heck and his family moved to California, where his interest in chemistry was sparked by his interest in orchids. He completed his undergraduate work at UCLA in 1952 and his PhD in 1954 under Saul Winstein. Next, he earned an NSF postdoctoral fellowship at the Swiss Federal Institute of Technology in Zurich with Prelog (who would go on to win the Nobel in 1975). In 1955, Heck went back to UCLA to work on the neighboring group effect, a phenomenon covered in almost all undergraduate organic texts now. In 1956, he took a position at Hercules Powder (which later became Ashland Inc.) where his supervisor, Dr. David Breslow, suggested Heck to do something with transition metals. In chatting about the Wacker Process with then colleague Pat Henry, the latter hypothesized that the intermediate palladium species decomposed via beta-elimination. Heck went on to explore what happens when there is no beta hydrogen atom to allow the beta-elimination and discovered the first of his palladium cross-coupling reactions. In 1971 he left Hercules for the University of Delaware and published in 1972 a more "user friendly" protocol for the reaction that now bears his name. Heck retired in 1989, meaning he was no longer an active researcher when winning the Nobel.

Ei-ichi Negishi

Ei-ichi Negishi was born in China as a Japanese citizen on July 14, 1935 and died in the U.S. on June 6, 2021. After moving to Harbin at the age of one, his family moved to Seoul, Korea, two years before the end of World War II and before the splitting of Korea. He enrolled in elementary school at six, a year ahead of normal. After the war ended, his family moved to Tokyo into a house that miraculously survived the bombing during the war. Negishi spent some of his formative years living on his family farm and spent large amounts of time playing with friends. Upon graduating after completing college in 1958 (his studies were extended a year because of a long illness), Negishi and his future wife announced their engagement to their parents, and he took a position at Teijin as a research chemist at the Iwakuni Research Laboratories. Negishi applied for and was awarded a three-year Fullbright Scholarship to pursue his PhD in the U.S. and in 1960, he started his studies at the University of Pennsylvania, eventually completing his PhD in 1963 under A. R. Day. While at Penn, Negishi got the opportunity to attend lectures and even talk to as many as ten prior or eventual Nobel Laureates. By his telling, these meetings made the Nobel Prize something of a reality to him; it may even be something he may one day have a shot at. Negishi then began his independent career at Syracuse University in 1972, eventually moving to Purdue in 1979. Negishi's comment about meeting Nobel Laureates is certainly part of why I include these bios as it hopefully comes close to humanizing the winners for those readers so unlucky as to never meet one in person.

Akira Suzuki

Akira Suzuki was born in Japan on September 12, 1930. Although in high school he was interested in math, in his first year at Hokkaido University, he read *Textbook of Organic Chemistry* by Fieser and Fieser and there was no looking back; Suzuki decided to major in organic chemistry. In 1959, he completed his PhD from Hokkaido University's Graduate School of Science, subsequently taking up a position as a research assistant. In 1961, he was invited to become Assistant Professor of Synthetic Organic Chemistry Lab at the newly founded Synthetic Chemical Engineering Department in the Faculty of Engineering. In 1973, he succeeded Professor H. Otsuka in the Applied Chemistry Department. Retiring in 2002, Suzuki has been awarded multiple honors in addition to the Nobel Prize. Like Heck, Suzuki was no longer active by the time of his award.

As this introduction, and in particular, the biographies section, closes, it is my hope that at least one of my readers feels some manner of personal connection and inspiration from any one of these stories. Every Nobel Prize winner of every discipline started somewhere, just like you, dear reader. Maybe one of you will win one someday too!

CITED REFERENCES

1   www.nobelprize.org/nomination/chemistry/, last checked 6/29/22.
2   www.nobelprize.org/nomination/physics/, last checked 6/29/22.

3   www.nobelprize.org/nomination/literature/, last checked 6/29/22.

4   www.nobelprize.org/nomination/literature/, last checked 6/29/22.

5   www.nobelprize.org/nomination/peace/, last checked 6/29/22.

6   www.nobelprize.org/nomination/economic-sciences/, last checked 6/29/22.

7   I must credit and thank James Leahy for alerting me to this story about Donald Cram.

8   www.nytimes.com/1987/10/15/us/caller-for-the-academy-gets-a-wrong-number.
    html, www.latimes.com/archives/la-xpm-1987-10-15-mn-14323-story.html, last
    checked 8/29/23.

9   I must credit and thank my wife, Kerry Kautzman for alerting me to this story.

10 www.loudersound.com/news/tom-morellos-ex-bandmate-has-been-awarded-the-
    nobel-prize-in-chemistry, https://twitter.com/tmorello/status/1577694126610337795,
    https://metalinjection.net/news/2022-nobel-prize-in-chemistry-winner-used-to-be-in-
    a-band-with-tom-morello, last checked 8/29/23.

11 I must credit and thank my colleague, Likin Simon-Romero for alerting me to
    this story.

12 www.snopes.com/fact-check/the-prizes-rite/, last checked 8/29/23.

13 www.chemistryworld.com/features/are-the-nobel-prizes-good-for-science/3009
    557.article, last checked 6/30/22.

14 www.nobelprize.org/nomination/archive/search.php, last checked 2/23/24.

15 https://cen.acs.org/articles/94/i15/Five-chemists-should-won-Nobel.html   last
    checked 2/23/24.

16 https://en.wikipedia.org/wiki/George_Washington_Carver, https://web.archive.org/
    web/20070930090502/http://www.time.com/time/magazine/article/0,9171,801
    330,00.html, last checked 8/29/23.

17 www.usnews.com/news/best-countries/articles/2020-10-01/the-nobel-prizes-have-
    a-diversity-problem-worse-than-the-scientific-fields-they-honor, last checked 6/30/
    22.

18 www.science.org/content/article/one-reason-men-often-sweep-nobels-few-
    women-nominees, last checked 6/30/22.

19 Borman, Stu; Chemical and Engineering News, September 11th, 2017, 22–24,
    "Women Overlooked for Nobel Honors"

20 www.popsci.com/racial-inequality-nobel-prize/, last checked 6/29/22.

21 www.bbc.com/news/world-europe-58875152, last checked 6/30/22.

22 https://massivesci.com/articles/nobel-prize-science-gender-physics/, last checked
    6/30/22.

23 *Science* **2022**, 399.

24 *JACS* **2022**, 4819.

25 www.cnn.com/2021/10/06/opinions/physics-nobel-climate-change-lincoln/index.
    html, last checked 6/29/22.

26 https://improbable.com/ig/winners/, last checked 9/21/2023.

GENERAL SOURCES

www.nobelprize.org/, last checked 6/29/22.

https://chemistry.as.miami.edu/_assets/pdf/murthy-group/gnl_jensen-2.pdf,   last checked 6/29/22.

2001
Sharpless, Knowles, and Noyori

2

One important aspect that the two shares of the 2001 Nobel Prize in chemistry have in common is that they pertain to stereoselective reactions. The 2001 Nobel Prize in chemistry was shared between three people for two topics. K. (Karl) Barry Sharpless was awarded half of the award for "his work on chirally catalyzed oxidation reactions" and the remaining half was split between William Knowles and Ryoji Noyori for "their work on chirally-catalyzed hydrogenation reactions". Previously, in 1975, John Warcup Cornforth and Vladimir Prelog were awarded the Nobel Prize in chemistry for "work on the stereochemistry of enzyme-catalyzed reactions" and "research into the stereo-chemistry of organic molecules and reactions", respectively. Since then, the 2001 award is the first given in relation to stereochemistry or stereoselective reactions.

This chapter is slightly different from the other two in this volume. While the other two, like this one, were awarded to three individuals, this award was given for two different topics whereas the other two were awarded for the same topic or at least related topics. For the sake of retaining a consistent format across this series' volumes, this year's awards are combined into one chapter. That both "sections" of this chapter are given for stereoselective reactions only further justifies this decision over two different chapters. The chapter will how-ever be broken down into two sections. The first section will be devoted to the work of Sharpless and will follow the format of first covering the discovery and then modern applications and/or modifications, and ending with noteworthy applications, the usual format for earlier volumes of this series. The second part of the chapter will cover the work of Knowles and Noyori, duplicating the format.

Before proceeding further, for the sake of the novice reader, a brief discus-sion of stereochemistry and its importance is warranted. Effectively, at its most simplistic reduction, stereochemistry can be considered the "handedness" of

DOI: 10.1201/9781003006879-2

a molecule, even the handedness of individual atoms within a molecule with many "hands." Rather than a left-handed and right-handed molecule, however, they are given the labels R and S. How this is determined and the origin of the R and S label are irrelevant for this discussion. For an example of stereochemical possibilities, consider the molecules in Figure 2.1.

In the case of **1a** and **1b** (Figure 2.1, box 1), these structures are nonsuperimposable mirror images and are referred to as enantiomers of each other. Meanwhile, **2a** and **2b** (Figure 2.1, box 2) are not mirror images and are referred to as diastereomers of each other. In **2a** and **2b**, notice that there is more than one atom that has handedness, called chirality. Such chiral atoms are referred to as chiral centers or asymmetric centers.

Stereochemistry often has a dramatic impact on how living organisms (like we humans) respond to the molecule. The structures in Figure 2.1, boxes 3 and 4 illustrate this well and in perhaps (in)famous ways; thalidomide (**3a** and **3b**, Figure 2.1 box 3) is a tragic example. The R enantiomer (**3a**) is a sedative that was formerly used to alleviate symptoms of morning sickness in pregnant women while the S enantiomer (**3b**) is a teratogen responsible for many birth defects in the late 1950s/early 1960s. A less traumatic example is carvone (**4a** and **4b**, Figure 2.1 box 4). The S isomer (**4a**) has a caraway or rye scent while the R isomer (**4b**) has a sweet mint smell. All of this to make the point that stereochemistry is important. In truth, an entire book could easily be written around such examples. Short of that here, chemical reactions that selectively create one stereoisomer in favor over the others are highly valued due to this importance of stereochemistry to biological activity/response. Control of the reaction products allows for control over biological activity, even if several steps downstream. Doing so selectively (or better yet specifically) allows for easier purification of products as well, reducing the cost of production especially on large scales.

Some important terminology housekeeping is necessary before exploring the reactions covered in this year's Nobel: stereoselectivity vs. stereospecificity. A stereoselective reaction simply favors the formation of one isomer over the other(s), while a stereospecific reaction generates only one isomer when more than one were possible. For simplicity's sake, a stereospecific reaction can be fairly considered a super, or *perfectly* stereoselective reaction. Reactions that selectively make one diastereomer over the other options are referred to diastereoselective/specific and in the case of enantiomers as enantioselective/specific; using the prefix *stereo* (rather than enantio- or dia-) encompasses both as a general label. Finally, the concept of % enantiomeric excess (%ee), how it is calculated, and how it is experimentally determined should be discussed.

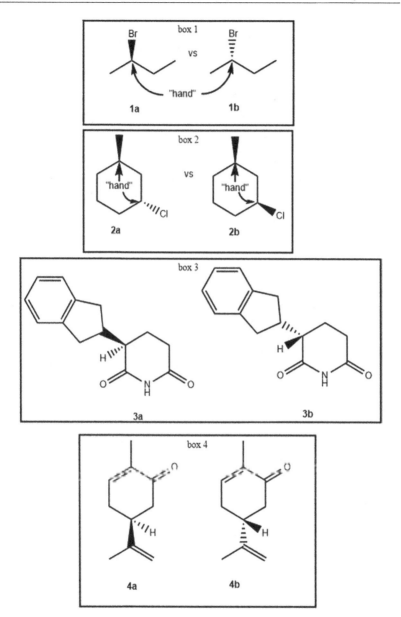

FIGURE 2.1 box 1 and 2: "Handedness" of molecules; box 1: enantiomers; box 2: diastereomers; box 3: thalidomide; box 4: carvone.

CALCULATING % ENANTIOMERIC EXCESS (%EE) AND EXPERIMENTALLY DETERMINING THE RELATIVE AMOUNTS IN A MIXTURE OF STEREOISOMERS

The amount of each enantiomer present in a mixture can be experimentally measured by at least two ways. The first is using the optical rotation of a solution of the sample. This approach only is viable if the optical rotation of *pure* enantiomer is known. For this purpose, either enantiomer will suffice. As an example, if it is known that pure *R* isomer has an optical rotation of −4.3, the *S* isomer will have an optical rotation of +4.3. They are always exactly equal magnitude but opposite sign. If the sample has an optical rotation of 0, this means there is an equal amount of each enantiomer, canceling each other out. If on the other hand, the value is nonzero, say +3.4, there is more of one than the other; in this hypothetical case, more *S* than *R*. If we assign the percent of *R* present to x and the percent of *S* present to y, the following mathematical equation is true since the magnitude of each isomer's contribution is directly proportional to its relative amount:

$3.4 = -4.3x + 4.3y$

Moreover, since the mixture is only made up of the *R* and *S* enantiomers in the solvent, we can also say that x + y = 1 (or 100%). Rearranging this equation so that x is redefined in terms of y tells us that x = 1−y. Using this definition of x, a new equation can be used:

$3.4 = -4.3(1-y) + 4.3y$

where 1−y has replaced x based upon this equivalence. Mathematically, this is important since now we only have one variable to solve. Solving for y as shown will give us the percent of the *S* isomer.

$-4.3 + 4.3y + 4.3y = 3.4$

$8.6y = 7.7$

$y = 0.895 \text{ or } 89.5\%$

$x = 100-89.5 = 10.5\% \text{ (or } 0.105)$

With the percentage of the S isomer being 89.5% and that of the R isomer being 10.5%, the %ee (enantiomeric *excess*) is the difference between the two or 79%; it is the amount that one of the enantiomers is in excess over the other.

When the optical rotation is unknown or when the relationship between the compounds being compared is diastereomers, optical rotation cannot be used. For this purpose, nuclear magnetic resonance (NMR) is most commonly employed, specifically proton NMR (HNMR). An additional step is necessary though for enantiomers—the use of a chiral additive of some sort that allows for the distinguishing of enantiomers. Enantiomers otherwise have *perfectly* overlapping signals in the NMR; diastereomers can *usually* be relied on to have at least one signal in the spectra that do not overlap. Whatever the case may be, the spectrum returned by the instrument will allow the user to quantify how much there is of one, relative to the other. The advent of modern computers makes this even easier in the sense that you can tell the computer that the smaller peak (the one coming from the compound in lesser quantity) is equal to a value of 1 and the computer will automatically compute for the user the value of the other peak, giving a ratio of sorts.

As another hypothetical example, imagine that the ratio of enantiomers or of diastereomers was found by HNMR to be 13.4:1. Now, the percent of each is easily calculated using the formula here:

$$\% \, stereoisomer \, 1 = \frac{13.4}{(13.4+1)} = 93.1\%$$

$$\% \, stereoisomer \, 2 = \frac{1}{(13.4+1)} = 6.9\%$$

If the pair in this second case were also enantiomers, the %ee would be 86.2 and if the pair were diastereomers, we would instead call it the % diastereomeric excess (or %de). With this terminology in hand, let's now start with the work of Barry Sharpless.

SHARPLESS'S WORK

While the discussion about stereochemistry and even parts of the regiochemistry one can be applied to the entirety of the 2001 Nobel Prize in Chemistry, Sharpless's share of the award was given not for one but two separate oxidation reactions: the stereoselective epoxidation of an alkene in an allylic alcohol

(Figure 2.2, box 1) and the asymmetric dihydroxylation of alkenes (Figure 2.2, box 2).

In the case of epoxidation, the presence of the alcohol (hydroxyl, OH group) in the allylic position (see the inset) is essential for the reaction. The diethyl tartrate (DET) additive is what drives the stereochemistry of the reaction. In the case of the L-(+)-DET (**5a**), it favors the "top" face of the alkene to selectively give epoxide **6a** when drawn in the orientation shown. Meanwhile, the D-(−)-DET (**5b**) favors delivery of the oxygen to the "bottom" face of the alkene to give epoxide **6b** when drawn in the orientation shown. In both cases, stereoselectivities in the range of 100:1 are commonly observed. In the reaction, the titanium reagent first coordinates with the DET derivative. The titanium atom then coordinates with the *tert-butyl hydroperoxide* (t-buOOH) before the hydroxyl group of the allyl alcohol coordinates with the titanium as well. This makes it clear why the process must employ allylic alcohols. This simultaneous coordination of the titanium with the allylic alcohol and the peroxide places these moieties in close proximity around the titanium. Because of the stereochemistry of the DET derivative, the oxidant (the peroxide) and the allylic alcohol can only encounter/approach each other from one direction. Sharpless's other work (Figure 2.2, box 2) involves stereoselective dihydroxylation of an alkene and is covered in greater detail later in this section and employs a cinchona alkaloid as a chiral agent (**7**).

Regarding asymmetric epoxidation, Sharpless points to geraniol (**8**) as an ideal scaffold for selectivity in epoxidation of alkenes since it has two equally substituted alkenes where one is allylic (Figure 2.3). The problem of epoxidizing geraniol, Sharpless notes, is twofold: regiochemistry (which double bond reacts) and stereochemistry (which direction the epoxide oxygen ultimately points in). Henbost[1] used peracids to solve the regiochemical challenge, selectively epoxidizing the isolated alkene. Almost two decades later, Michaelson and Sharpless solved selectively epoxidizing the allylic alkene using transition metals.[2] Katsuki and Sharpless,[3] provided the last piece of the puzzle, performing the first of the asymmetric epoxidations. This finally solved the enantioselective epoxidation of the alkene with the hydroxyl group (OH) in the allylic position. After this, Jacobsen[4] and Katsuki[5] independently developed methods of asymmetrically epoxidizing isolated alkenes, a reaction discussed briefly later in this chapter in the section about modern and further developments.

Some of the key features of the Sharpless Asymmetric Epoxidation (SAE) reaction, as it is now called, are its simplicity and the use of inexpensive, easy to obtain reagents. The predictability of the reaction's stereochemistry (see Figure 2.2, box 1) is observed regardless of substitution pattern of the alkene. Although for non-reactive alkenes, a full equivalent of both titanium isopropoxide and diethyl tartrate are needed; for more reactive allylic alcohols,

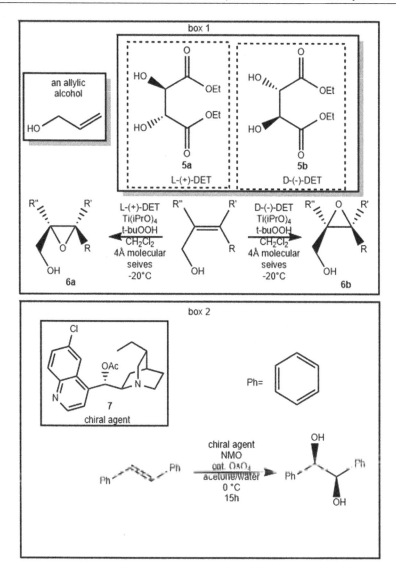

FIGURE 2.2   Sharpless asymmetric epoxidation (box 1) and Sharpless asymmetric dihydroxylation (box 2).

a catalytic amount of each is enough, making this transformation remarkable. One noteworthy complication is difficult extraction and isolation of products if the epoxy alcohol has a high water solubility. This—for the sake of the novice reader—can be said about any chemical reaction in the field of organic chemistry. Water-soluble products are often very difficult to isolate. Such points are

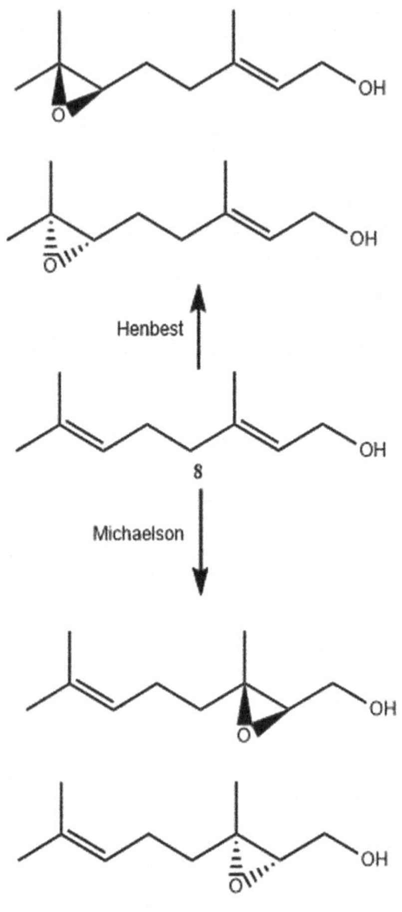

FIGURE 2.3 Epoxidations of geraniol.

important to keep in mind when planning a synthesis, as important as the cost and relative safety of reagents. They are also what separate chemical reactions on paper form *real* lab work.

Sharpless Asymmetric Dihydroxylation

Sharpless's share of the award was given for not one but two reactions: the aforementioned SAE and Sharpless Asymmetric Dihydroxylation (SAD). Sharpless and his lab discovered SAD in two stages over nearly a decade. The first stage, reported in 1980, established that it was possible to perform the dihydroxylation of an alkene in a stereoselective manner, using a stoichiometric amount of OsO$_4$. Due to the high toxicity and cost of osmium, this was deemed undesirable, a reasonable designation to assign. To achieve the stereoselectivity, Sharpless first turned to a chiral pyridine derivative (**9**, Figure 2.4). Pyridine was chosen because it was known to increase the rate of reactions of OsO$_4$ with alkenes. Sharpless reasoned that if the pyridine was coordinating with the metal center to accelerate the rate, a nearby chiral center might induce selectivity in the formation of the product by influencing the direction that the reacting partners can approach each other. Although an influence was observed, it only brought about modest ee values of 3%–18%. Griffith had shown that **10** forms more stable complexes with OsO$_4$,[6] leading Sharpless to pursue **11** in place of pyridine as the next step. It was reasoned that with the chiral center immediately neighboring the nitrogen atom called out in Figure 2.4, the influence and thus the %ee would be better and this was found to be true.

Stage two was reported in 1988 when a catalytic procedure was demonstrated. Here, Sharpless and coworkers[7] incorporated N-Morpholine oxide (NMO), according to the procedure developed by Upjohn.[8] NMO re-oxidizes the osmonate by-product to recreate the OsO$_4$ reagent *in-situ*. With the active ingredient (OsO$_4$) being remade under the reaction conditions, only

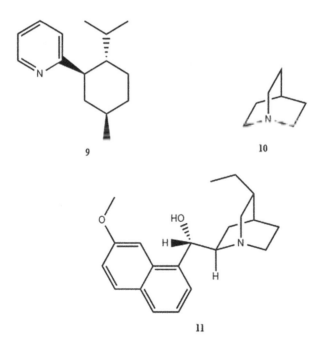

FIGURE 2.4 Chiral additives for asymmetric dihydroxylations.

a catalytic amount of this expensive and toxic reagent is necessary. The magnitude of this breakthrough is difficult to overstate.

MODERN DEVELOPMENTS

Since Sharpless's report of what is now known as SAE, Jacobsen[4] (**12a**, Figure 2.5) and Katsuki[5] (**12b**, Figure 2.5) independently reported a method using very similar reagents, that extends the process to isolated alkenes. Each employed a manganese catalyst as the catalytic oxidant and chiral director and iodosyl mesitylene as the oxygen source. Overall, the percent yields and %ees observed in Jacobsen's work were better than Katsuki's. Altogether, these reactions have tremendous synthetic value. Epoxides especially can be further functionalized a large number of ways with regioselectivity and stereospecificity.

With the combined (Jacobsen, Katsuki, and of course Sharpless), asymmetric epoxidation work, epoxides can be generated stereoselectively, making the stereo*specificity* of the epoxide reacting more able to make desired isomers in very high yields. This is because when an epoxide reacts, it "opens" and the incoming group *always* approaches the carbon atom from the opposite direction the epoxide is pointing.

The asymmetric dihydroxylation has similarly been improved upon to higher levels of enantioselective control and improved rates of reaction. Beller

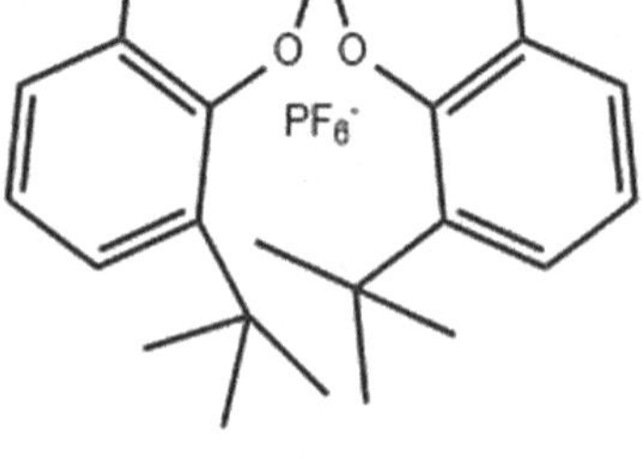

12a 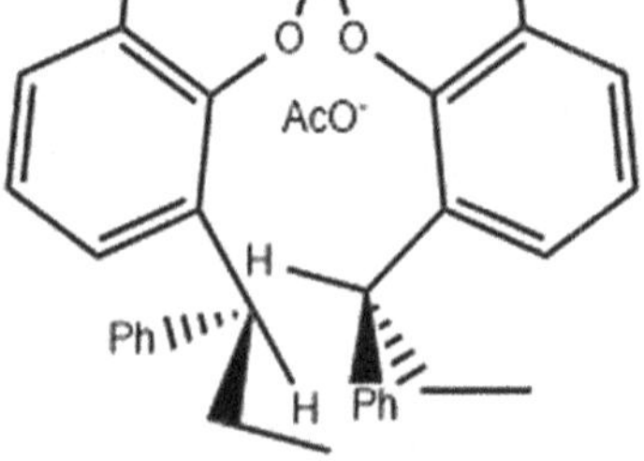   **12b**

FIGURE 2.5   Jacobsen's (**12a**) and Katsuki's (**12b**) asymmetric epoxidation catalysts.

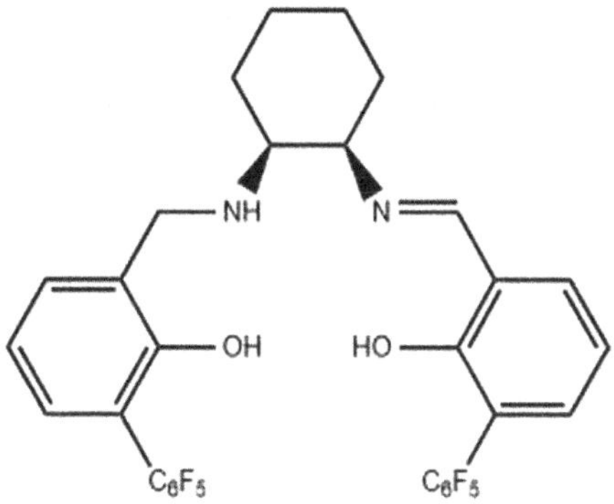

13

FIGURE 2.6   Berkessel's ligand for syn epoxidations.

and coworkers reported in 2000[9] that internal olefins react at a constant pH of 12.0 at an improved reaction rate, along with the omission of the hydrolysis aids. Terminal olefins at pH levels of 10.0 at room temperature meanwhile lead to higher %ees. The improvements here were modest at best, in most cases though one saw 33% ee rise to 61% ee.

Even more recently, another modification was reported by Berkessel and coworkers.[10] Typically, SAE gives the anti-product relative to the OH epoxide. By using the diphenolic ligand **13** (Figure 2.6), they were able to achieve equally high preference for the syn formation of the epoxide in very high yields.

APPLICATIONS TO SYNTHESIS OF NATURAL PRODUCTS AND PHARMACEUTICAL COMPOUNDS

Both reactions pioneered by Sharpless have seen widespread use in the synthesis of natural products and pharmaceutical compounds. This should be no surprise. Not only are oxygen atoms ubiquitous in both, it is also commonly present in stereospecific orientations. That these reactions stereoselectively deliver oxygen atoms to molecules makes them a very popular, even necessary, choice indeed. Examples of the use of both reactions (SAE and SAD) and the closely related Sharpless Asymmetric Aminohydroxylation (SAA) reaction (which was not part of the Nobel-winning work) abound but in the interest of time and space, only a handful of cases are considered. These topics have been reviewed, for example Muñoz and coworkers wrote a recent review of SAE that focuses on applications in the synthesis of bioactive natural products.[11] Heravi and Coworkers also reviewed SAE in total synthesis.[12]

One example in their review is Lan et al.'s[13] synthesis of (+)-11,12-epoxysarcophytol A (**14**, Figure 2.7), a compound that has shown promise in cancer prevention. Starting with the E, E-acetylfarnesol derivative, the target was prepared in 42% overall yield. SAE was done on the farnesol derivative to give the desired epoxide to give the epoxide **14**.

Clavilactone B (**15**, Figure 2.7), one of a family of compounds found to be potent inhibitors of EGFR tyrosine kinase[14–16] was first synthesized by Larrosa et al.[15] Starting with propargyl alcohol with one of the key steps being a high-yielding SAE to give epoxide from the allylic alcohol. EGFR tyrosine kinase is responsible for a number of highly important signal transduction pathways such as cell growth, differentiation, and apoptosis.

Extracts of the plant *Cleistochlamys kirkii*, found in Tanzania and Mozambique, belonging to the Annonaceae family have been used as a traditional medicine to remedy skin infections, tuberculosis, and rheumatism.[17] Modern science has identified (−)-cleistenolide (**16**, Figure 2.7) as a component possessing activity against *Staphylococcus aureas*, *Bacillus anthacis*, and *Candida albicans*.[18,19] Schmidt et al.[20] completed the first total synthesis of this compound using two Nobel Prize-winning works. The first was SAE to set the indicated stereocenter. Subsequent steps used ring closing metathesis, covered later in this volume.

SAE was used by Yokoyama et al.[21] in the synthesis of (+)-dienomycin c (**17**, Figure 2.7). Dienomycin C has antibacterial activity, including against tuberculosis.

The synthesis of radicamine B[22] (**18**, Figure 2.7) was achieved using SAE as a key step. The radicamines are polyhydroxylated pyrrolidine alkaloids isolated from *Lobelia chinesis*. The SAE was done on the cinnamyl alcohol derivative with further elaboration yielding the target.

Lactacystin (**19**, Figure 2.7), which has promise in the treatment of arthritis, asthma, and Alzheimer's disease, was synthesized by Brennan and coworkers[23] using SAE as a key step. Even in the presence of a typically more reactive alkyne, the SAE reaction works at the alkene allylic to the alcohol to give the epoxide in very good yield.

Researchers at the Auckland Cancer Research Centre, Advinus Therapeutics, and the Drugs for Neglected Disease Institute used SAE in a scalable process for the synthesis of a drug candidate (**20**, Figure 2.7) to treat visceral leishmaniasis.[24,25] Leishmaniasis is an extremely deadly disease in Asia, East Africa, South America, and the Mediterranean,[26] spread by the bite of phlebotomine sand flies. Cutaneous leishmaniasis causes skin sores while visceral affects internal organs such as the liver, spleen and bone marrow.[27]

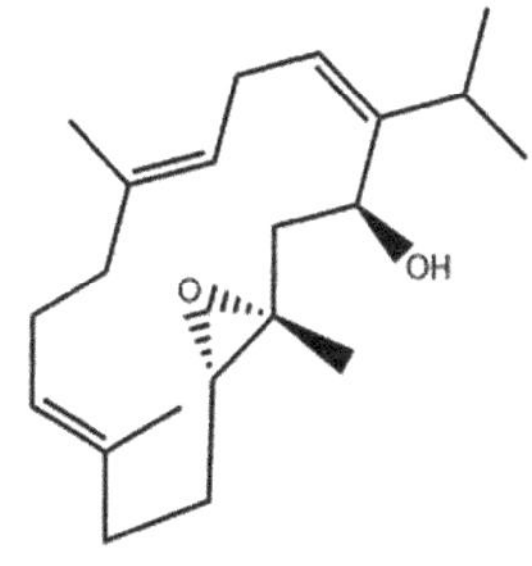

14

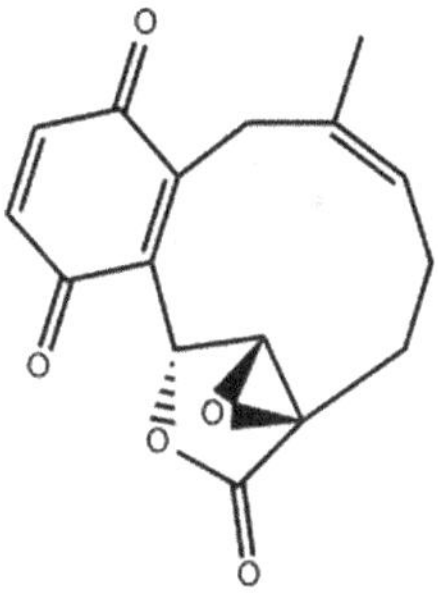

15

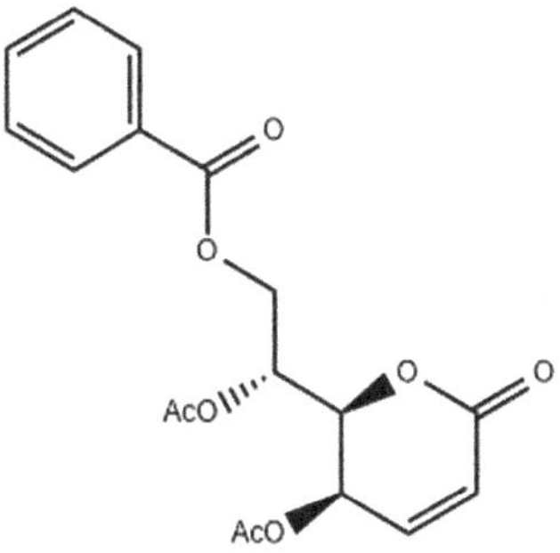

16

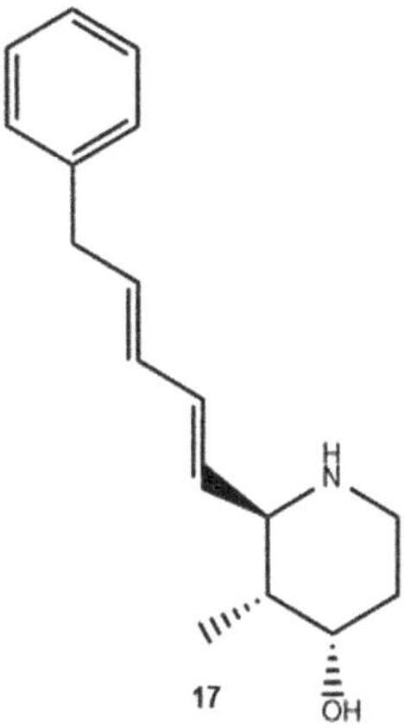

17

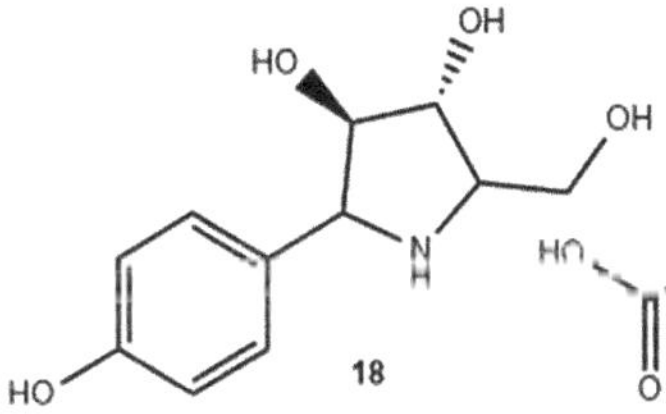

18

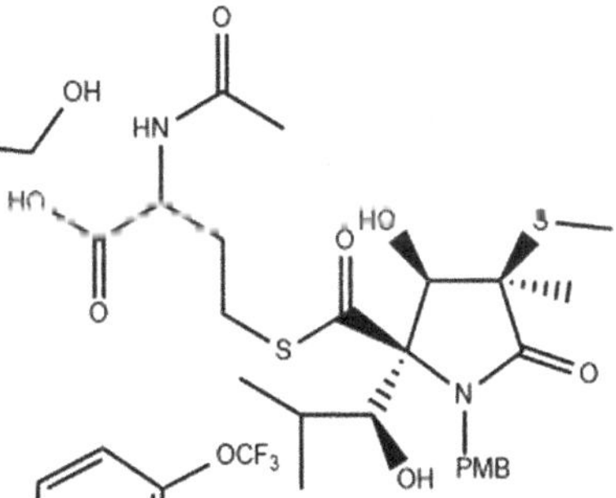

19

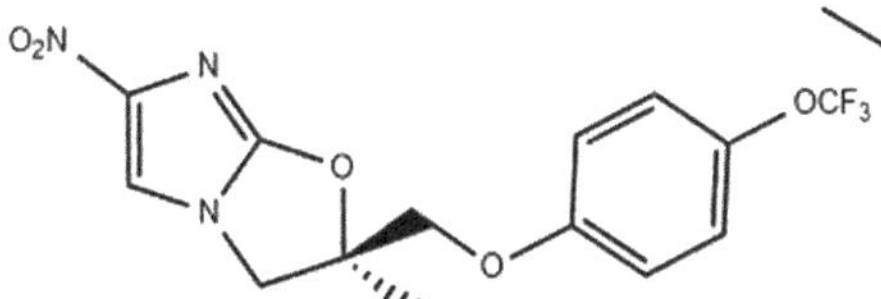

20

FIGURE 2.7 Natural products and interesting pharmaceutical compounds where Sharpless asymmetric epoxidation played a role in the synthesis.

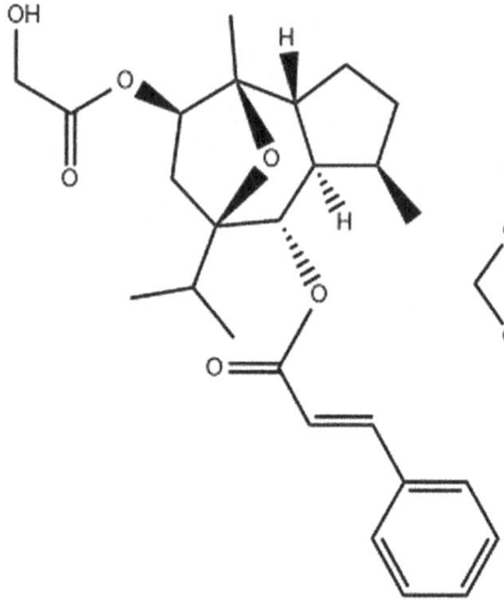

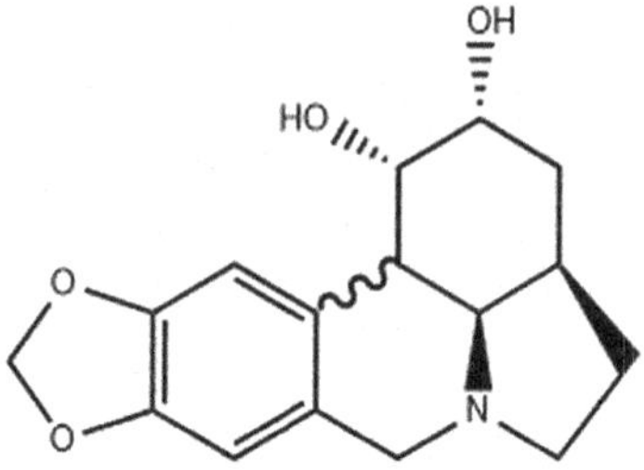

FIGURE 2.8 · Natural products and interesting pharmaceutical compounds where Sharpless asymmetric dihydroxylation played a role in the synthesis.

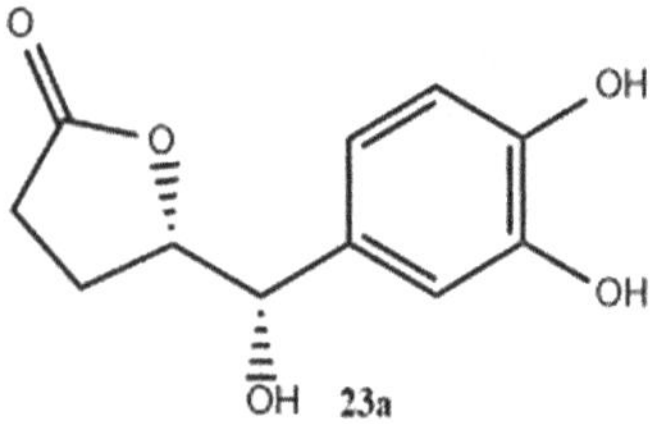

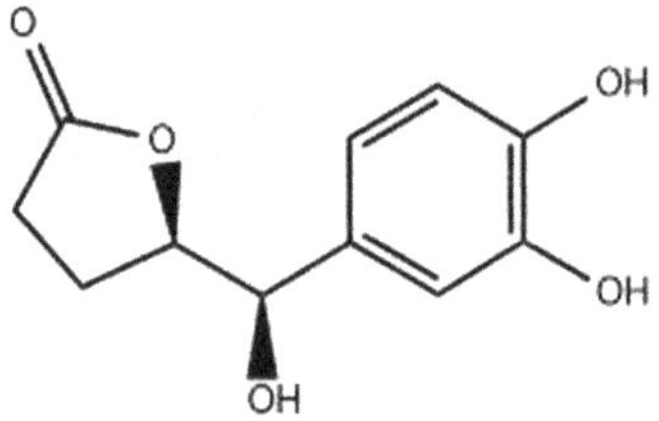

Sharpless Asymmetric Dihydroxlation in Natural Product Synthesis

SAD has also seen widespread use in the synthesis of natural products and has been reviewed, for example by Mojzych and coworkers.[28] (−)-Zephyranthine (**21**, Figure 2.8) belongs to a class of alkaloids called lycorines. These compounds are known to have important potential medicinal uses[29] as they interrupt acetylcholine and uncontrolled division of cells.[30] Zhao and coworkers used SAD to make the compound in 2021.

Englerin (**22**, Figure 2.8), isolated from *Phyllanthus engleri* was found to have activity against kidney cancer cells[31] and activates protein kinase, allowing it to regulate glucose levels and was synthesized by Mou and coworkers in

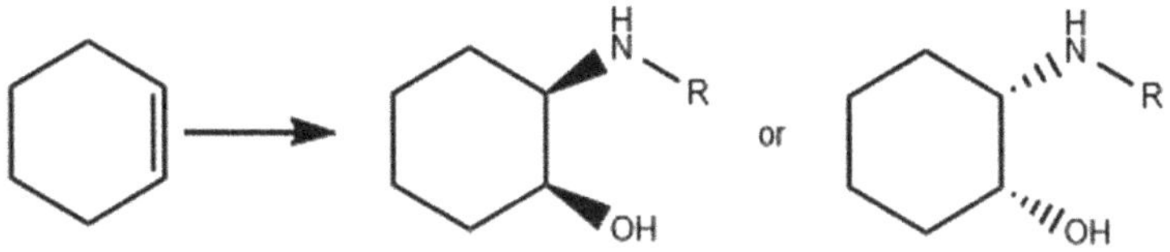

FIGURE 2.9   Sharpless asymmetric aminohydroxylation.

2020.[32] In their approach, the SAD occurs in tandem with opening an epoxide to generate the furan ring in the product.

Both enantiomers (**23a, 23b**, Figure 2.8) of the anti-inflammatory DHPV were synthesized by Kim and coworkers[33] in 2020. These lactones have been extracted from the intestines of vertebrates and are believed to come from microorganisms that live inside the digestive tracts.[34] The dihydroxylation is immediately followed by the lactonization reaction. Subsequent deprotection of the alcohol (phenol) gives the final target.

Another contribution by Sharpless that, although related to the asymmetric dihydroxylation, was not recognized by this award, nor his 2022 award, is known as Sharpless asymmetric aminohydroxylation (Figure 2.9). This reaction results in the formation of, rather than a cis diol as in the case of SAD, a cis amino alcohol, meaning that that alcohol and amine (as a protected amine) are pointing in the same 3D direction. Although not covered in more detail here, the general case is shown in Figure 2.10 and its use in the synthesis of natural products has unsurprisingly been reviewed like the others.[35] Note the similarity to SAD. Here, however, the result is a vicinal (on neighboring carbon atoms) difunctional molecule.

Asymmetric Hydrogenations: Knowles and Noyori

The other half of the 2001 Nobel Prize was evenly split between William Knowles and Ryoji Noyori for "their work on chirally catalyzed hydrogenation reactions." Knowles was the first to achieve the remarkable feat in 1968.[36] I call it remarkable because hydrogen (be it hydride (H⁻), a proton (H⁺), or H_2) is as small as it gets as far as chemical reagents go. As a result, they can fit just about anywhere, even tight, sterically crowded places. This makes it very difficult to force a bias for one direction of approach over the other(s). Knowles's work was impacted by two prior breakthroughs. The rhodium complex commonly called Wilkinson's catalyst—a homogeneous catalyst for hydrogenation reactions ($RhCl(PPh_3)_3$)—came first, courtesy of Osborn and Wilkinson.[37] Next was synthetic methods that established methods for the synthesis of optically active (chiral) phosphines, thanks to Horner[38] and Mislow.[39]

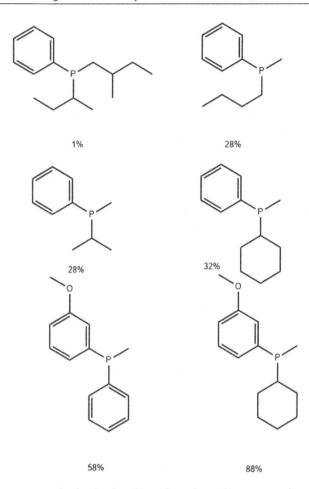

1%   28%

28%   32%

58%   88%

FIGURE 2.10 Initial chiral phosphine ligands with %ee employed during hydrogenation reactions.

Knowles had the critical insight of replacing the achiral (not chiral) PPh$_3$ in Wilkinson's catalyst with (−)-methylpropylphtenylphosphine that itself had 69% ee. Horner,[40–42] Kagan,[43] and Morrison and Bostnich[44] all followed up with similar results. The early results, (Figure 2.10) however, yielded only modest %ee in their hydrogenation reactions, far too modest to be synthetically useful. A major breakthrough came with the synthesis of L-DOPA (**24**, Figure 2.11) by Knowles et al.[45] L-DOPA is useful in treating Parkinson's disease. Their synthesis used DiPAMP as the catalyst (**25**, Figure 2.12), a chelating diphosphine with two chiral phosphorus atoms and 1,5-cyclooctadiene chelated to a cationic

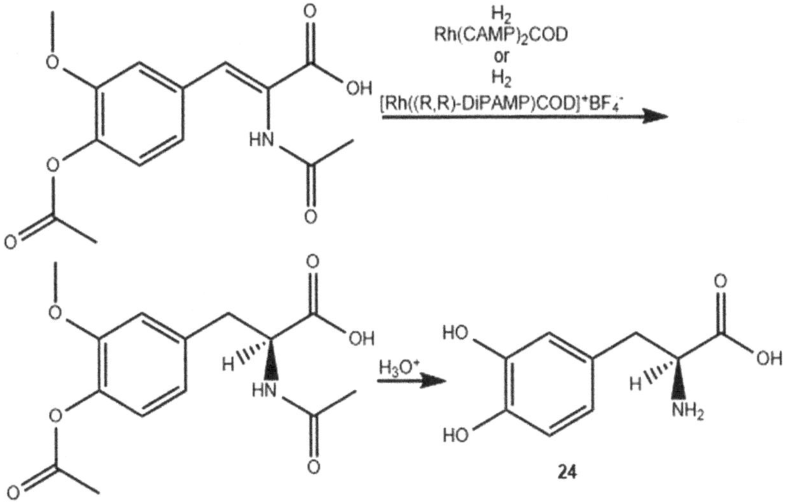

FIGURE 2.11 Synthesis of L-DOPA.

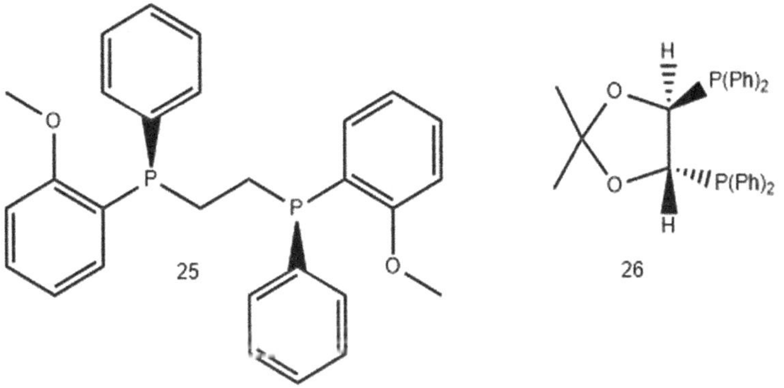

FIGURE 2.12 More efficient chiral phosphine ligands.

Rh. This system has the important and noteworthy restriction of substrate specificity limited to enamines.

Some of Knowles's work also employed chiral alkyl side chains on the phosphorus. This provided barely detectable increases in the %ee of hydrogenation reactions. In part because the alkene and phosphorus must coordinate to the Rh at the same time in the mechanism. The decision was made to make the phosphorus the chiral atom, placing any stereochemical bias closer to the metal coordination site. Subsequently, Kagan prepared diphosphines such as R,R-DIOP (**26**, Figure 2.12), based on tartaric acids, that placed a chiral bias on carbon achieving similar %ees. This started a shift to biphosphine ligands

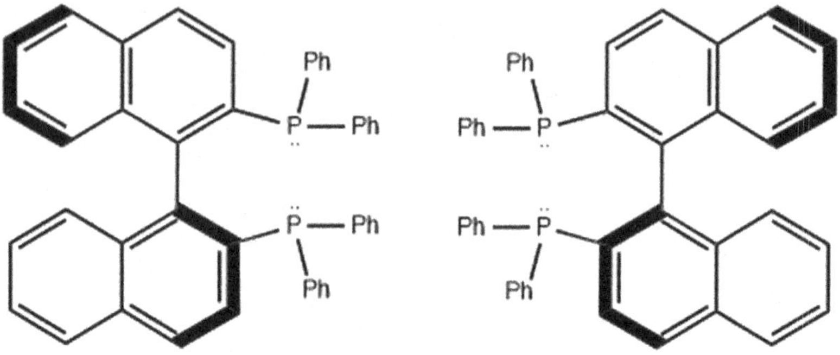

27a

27b

FIGURE 2.13   Structure of S and R BINAP.

with a chiral carbon backbone. Knowles followed this up with a biphosphine of his own—R,R-DiPAMP (**25**)—which gave 95% ee with their L-DOPA synthesis, even better than anything before.

One drawback noted by Knowles in his Nobel lecture is the high cost of preparing the chiral phosphines though in the next moment during his lecture, he highlights that even a very small amount will yield huge quantities of chiral product. This allows for a tremendous return on the investment of preparing or purchasing these reagents.

Noyori first became interested in the asymmetric hydrogenation while a postdoctoral research associate in E. J. Corey's lab (Nobel Prize 1990, Volume 2 of this series) during which Noyori met a former associate of Wilkinson and co-inventor of Wilkinson's catalyst, J. A. Osborn. At that point, in the early 1970s, only dehydroamino acids were known to submit to hydrogenation catalysis by chiral phosphine-chelated Rh complexes. Other olefins were still untapped and ketones were likewise unknown. Noyori's contribution was assisted by the elucidation of the mechanism by Halpern.[46] Through Halpern's work, it became clear that the larger %ees needed for industrial (and really *all* synthetic processes) would be achievable if the diastereomeric-activated complexes that are intermediates in the reaction had a greater difference in relative energy. Noyori and Takaya[47] discovered BINAP, a chiral diphosphine (Figure 2.13)—preparable in both the *S* (**27a**) and *R* (**27b**) enantiomeric form—Rh(I) complexes that are extremely efficient at a variety of asymmetric catalysis including the isomerization of allylic amines to enamines and the enantioselective hydrogenation of α-(acylamino) acrylic acids and esters. The next major breakthrough came with Noyori's[48] discovery of BINAP-Ru(II) complexes. Here, Noyori found these catalysts to be very effective catalysts for the enantioselective hydrogenation of both α-, β- and β-, γ-unsaturated carboxylic acids. Using halogen-containing complexes, ketones can be smoothly reduced to secondary alcohols, even in the presence of other carbonyl-containing groups such as esters[49] in very high yields and enantiomeric purity. Noyori also found that addition of bases such as KOH and ethylenediamine cause hydrogenation to dramatically favor hydrogenation of ketones over the usually favored olefin (alkene), solving a long-standing challenge in synthesis. The ketone or aldehyde can be isolated or conjugated with alkenes, and still submit to reduction by hydrogenation, preserving the alkene.

Noyori's lab developed a convenient way to make, with the help of a resolution step, enantiomerically pure (S)- and (R)-BINAP.[5051] The key break-through came in 1986 when Noyori and coworkers designed the BINAP Ru[2+] complexes,[4852–54] unlocking the asymmetric hydrogenation of olefins once and for all. Though initial results gave %ees in the 80% range, perseverance led to %ees approaching 100% using cationic Rh species chelated with BINAP and norbornadiene.[5556] This system was immediately applied to the synthesis (Figure 2.14) of menthol (**28**)[5758] where the Rh BINAP complex was used to isomerize the allylic amine intermediate asymmetrically. The system was subsequently applied to the asymmetric hydrogenation of olefins.

Use of halogen-containing Ru(II) complexes then expanded the substrates to a range of functionalized ketones. Finally, while director of the ERATO molecular catalysis project in 1995, Noyori and coworkers achieved the asymmetric hydrogenation of simple ketones using catalysts of the general form RuCl$_2$(chiral diphosphine) (Chiral diamine). This catalyst system proved selective for the carbonyl (the C=O), even in the presence of olefins.

Modern Developments

Stereoselective hydrogenation has continued to evolve as a field. Like many other processes, one of the efficiencies developed since the original discoveries is the binding of the chiral agents to a polymer and aqueous protocols.

Supported noble metal catalysts in the asymmetric hydrogenation of C=O and C=C bonds has been reviewed, for example by Baiker and Meemkem.[59] In the supported-reagent arena, both the catalyst and the ligand have been the reagent on the support. For example, platinum catalysts have been bound to silica for stereoselective hydrogenation reactions.[60] Ketones have been asymmetrically hydrogenated using a ruthenium catalyst ligated with polymer-supported chiral 1,2-diphenylethylenediamine[61] and aromatic ketones have been hydrogenated asymmetrically using a polymeric catalyst made from a 1,2-diamine that is polymer-bound.[62] Also, polymer-supported monodentate chiral phosphite ligands have been employed in the asymmetric hydrogenation

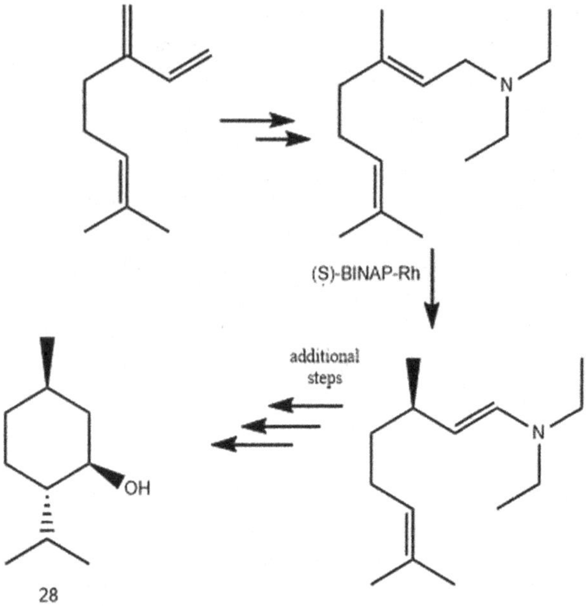

FIGURE 2.14   Synthesis of menthol.

of olefines.[63] Another example is a 2009 report[64] of rhodium-substituted carbonic anhydrase. Here, using a Rh(I) catalyst bound to carbonic anhydrase, *cis*-stilbene was shown to yield to hydrogenation over *trans*-stilbene in a 20:1 ratio. This is arguable different from the asymmetric hydrogenation discussed earlier in that it opens up enzyme-level selectivities for a substrate based on the substrates stereochemistry.

Water-compatible systems have also been described such as the carbohydrate-functionalized pyridincarboxyamide-bearing chiral iridium catalysts that were shown to asymmetrically hydrogenate α-keto acids.[65] Aqueous media asymmetric hydrogenation of ketones was also done using single-chain chiral Ru-Cu star (so called because of their shape) polymers.[66]

Use in Total Synthesis

Like any other chemical transformation of high importance, the use of asymmetric hydrogenation has been reviewed countless times, including but not limited to its use for natural products,[67–69] pharmaceutical compounds,[67] fragrance compounds,[67] and agrochemicals.[67] Tandem processes that combine stereoselective hydrogenation with other processes such as isomerization, oxidation, and epimerization have also been explored.[69] There are many examples of asymmetric hydrogenation being used in the synthesis of important targets such as natural products and compounds of pharmaceutical interest; too many to cover them comprehensively here. Included in this chapter is a brief discussion of some found in some of the reviews cited earlier and elsewhere.

One example is the preparation of chiral 2-aryl glycines (**29**) from N-aryl imino esters using nickel-catalyzed asymmetric hydrogenation (Figure 2.15).[70] Using the diphosphine chiral ligand in (2, 2, 2)-trifluoroethanol solvent, 99% conversion with 96% ee was observed in the model system. Such non-natural chiral amino acids have tremendous potential use in the synthesis of natural products, drugs, and especially synthetic peptides.

Chiral morpholines are likewise important structural motifs in biologically relevant compounds. These have also been synthesized using asymmetric hydrogenation.[71] Here, asymmetric hydrogenation led to easier access to precursors that were shown by the work of others to lead to the enantiomer of an inhibitor of GASK-3β[72] (**30**) and a D3 receptor agonist (**31**) (Figure 2.15).[73]

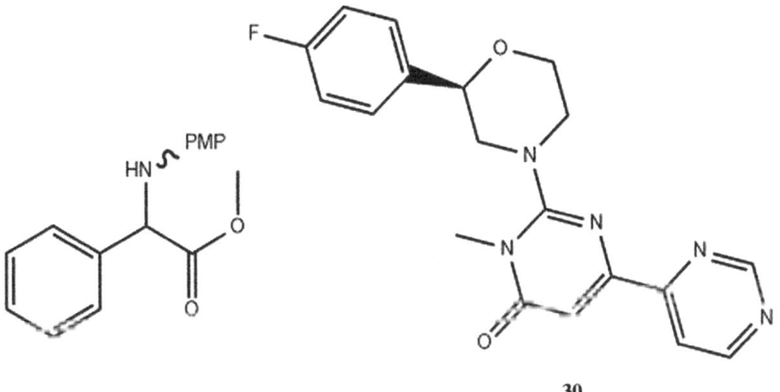

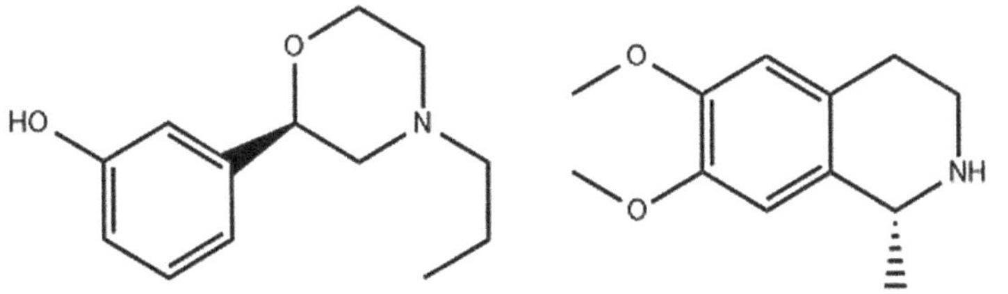

FIGURE 2.15 Natural products and interesting pharmaceutical compounds where asymmetric hydrogenation played a role in the synthesis.

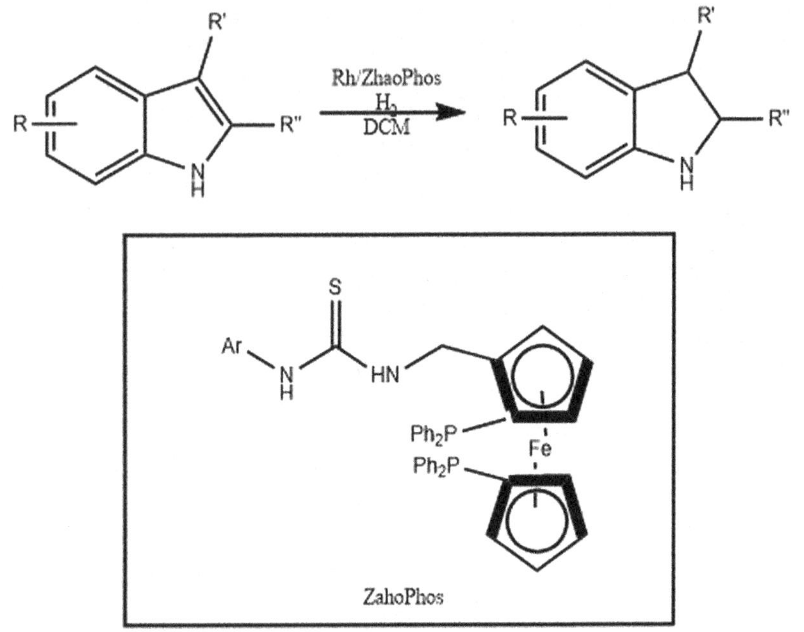

FIGURE 2.16   Hydrogenation of unprotected indoles.

Enamines have also been asymmetrically hydrogenated en route to natural products, for example, Opatz's synthesis of (+)-salsolidine.[74] The enamine is hydrogenated selectively in 91% ee to access this natural product (Figure 2.16).

Wen et al. reported[75] the asymmetric hydrogenation of *unprotected* indoles in 97+% ee on very large scales. The catalyst used in this work (Figure 2.16) was a Rh/ZhaoPhos catalyst.

SUMMARY

Asymmetrically selective or specific reactions are powerful in part because they mimic the selectivity usually observed within biological systems. Unlocking the power to perform such selective chemistry in the laboratory enables synthetic chemists to approach such selectivity when synthesizing products. That these hydroxylation and epoxidation reactions incorporate heteroatoms into molecules with such selectivity only makes them that much more powerful. With asymmetric hydrogenations, even chiral centers without a heteroatom can potentially be set selectively.

REFERENCES

(1)   *Journal of the Chemical Society* **1957**, 1958.
(2)   *Journal of the American Chemical Society* **1973**, 6136.
(3)   *Journal of the American Chemical Society* **1980**, 5974.
(4)   *Journal of the American Chemical Society* **1990**, 2801.
(5)   *Tetrahedron Letters* **1990**, 7345.
(6)   *Journal of the Chemical Society Dalton Transactions* **1977**, 941.
(7)   *Journal of the American Chemical Society* **1988**, 1968.
(8)   *Tetrahedron Letters* **1976**, 1973.
(9)   *Tetrahedron Letters* **2000**, 8083.
(10) *Angewandte Chimie International Edition* **2002**, e202201790.
(11) *Mini-Reviews in Organic Chemistry* **2021**, 606.
(12) *Tetrahedron Assymmetry* **2015**, 405.
(13) *Tetrahedron Letters* **2000**, 2181.
(14) *Molecules* **2010**, 1041.
(15) *Journal of the American Chemical Society* **2006**, 14042.
(16) *Organic Letters* **2015**, 126.
(17) *Tetrahedron Assymmetry* **2016**, 788.
(18) *Natural Products Communications* **2007**, 737.
(19) *European Journal of Organic Chemistry* **2014**, 8049.
(20) *Journal of Organic Chemistry* **2010**, 2389.
(21) *Tetrahedron* **2010**, 8458.
(22) *Tetrahedron Letters* **2011**, 4885.
(23) *Tetrahedron Assymmetry* **2015**, 405.
(24) *Journal of Tropical Mydicine Hygine* **2013**, 778.
(25) *Organic Processes Research and Development* **2016**, 52.
(26) *Andrews' Diseases of the Skin: Clinical Dermatology*; W. B. Saudners Co., 2006.
(27) www.who.int/news-room/tact-sheets/detail/leishmaniasis, last checked 8/10/24.
(28) *Molecules* **2023**, 2722.
(29) *Phytochemistry* **1985**, 2141.
(30) *Anti-Cancer Drug Design* **2000**, 1543.
(31) *Organic Letters* **2009**, 57.
(32) *Organic Letters* **2020**, 1976.
(33) *Molecules* **2020**, 2215.
(34) *Chirality* **2010**, 726.
(35) *RSC Advances* **2018**, 6634.
(36) *Chemical Communications* **1968**, 1945.
(37) *Journal of the Chemical society A* **1966**, 1711.
(38) *Tetrahedron Letters* **1961**, 161.
(39) *Journal of the American Chemical Society* **1967**, 4784.
(40) *Angewandte Chimie* **1968**, 1034.

(41)   *Angewandte Chimie International Edition* **1968**, 942.
(42)   *Tetrahedron Letters* **1968**, 4023.
(43)   *Chemical Communications* **197**, 481.
(44)   *Journal of the American Chemical Society* **1971**, 1301.
(45)   *Accounts of Chemical Research* **1983**, 106.
(46)   In *Asymmetric Synthesis Volume 5, Chapter 2*; Academic Press: NY, 1985.
(47)   *Journal of the American Chemical Society* **1980**, 7932.
(48)   *Inorganic Chemistry* **1988**, 566.
(49)   *Journal of the American Chemical Society* **1988**, 629.
(50)   *Journal of Organic Chemistry* **1986**, 629.
(51)   *Organic Synthesis* **1998**, 6.
(52)   *Journal of the American Chemical Society* **1986**, 7117.
(53)   *Journal of Organic Chemistry* **1992**, 4053.
(54)   *Organic Synthesis* **1993**, 74.
(55)   *Acta Crystallographica Section B* **1982**, 807.
(56)   *Journal of the American Chemical Society* **1980**, 629.
(57)   Tokyo Kagaka Dozin: Tokyo, 1992, p. 75.
(58)   Whiley Chichester, 1992.
(59)   *Chemical Reviews* **2017**.
(60)   *Applied Catalysis A: General* **2007**, 1.
(61)   *Organic Letters* **2003**, 4559.
(62)   *Journal of Organometallic Chemistry* **2007**, 487.
(63)   *Tetrahedron Letters* **2006**, 4263.
(64)   *Chemistry: A European Journal* **2009**, 1370.
(65)   *Organometallics* **2023**, 157.
(66)   *Journal of Catalysis* **2022**, 84.
(67)   *Chirality* **2023**, 121.
(68)   *Green Synthesis and Catalysis* **2020**, 26.
(69)   *Journal of the American Chemical Society* **2021**, 6724.
(70)   *Nature Communications* **2020**, 5835.
(71)   *Chemical Science* **2021**.
(72)   *Bioorganic Medicinal Chemistry Letters* **2012**, 6933.
(73)   *Bioorganic Medicinal Chemistry Letters* **2007**, 6991.
(74)   *European Journal of Organic Chemistry* **2007**, 3911.
(75)   *Organic Letters* **2018**, 2143.

2005
Chauvin, Schrock, and Grubbs

3

The 2005 Nobel Prize in chemistry was split in equal parts to three people: Yves Chauvin, Robert Grubbs, and Richard Schrock. All three received their prize "for the development of the metathesis method in organic synthesis." Chauvin's share of the prize was awarded for the elucidation of the mechanism—a feat that directly lead to the development of the catalysts of Schrock and Grubbs—while Grubbs and Schrock received their share for the development of highly efficient catalysts for the metathesis reactions. Of the metathesis reactions (Figure 3.1), arguably the most important is those called Ring-Closing Metathesis (RCM) reactions, so called because they are metathesis reactions that are intramolecular and so generate (i.e., close to form) a ring. This reaction is one of the most reliable and high-yielding ways to make medium-sized and large rings; previously such rings were more challenging to incorporate into synthetic targets. This power is, without question, part of why these combined breakthroughs merit a Nobel Prize.

WHAT IS METATHESIS?

Although there are a variety of reactions that are referred to as metathesis reactions, those considered here all employ starting materials that contain a carbon–carbon pi bond, an olefin/alkene. The reaction can occur between two alkenes, two alkynes, or one of each—an alkene and an alkyne. There are some specific catalysts that are more optimized or "tuned" for one optimal activity in specific cases with the passage of time. As discussed shortly, in some cases, these reactions are special in that they often permit access to synthetic targets in very high yields in cases where little to no other viable options exist.

DOI: 10.1201/9781003006879-3

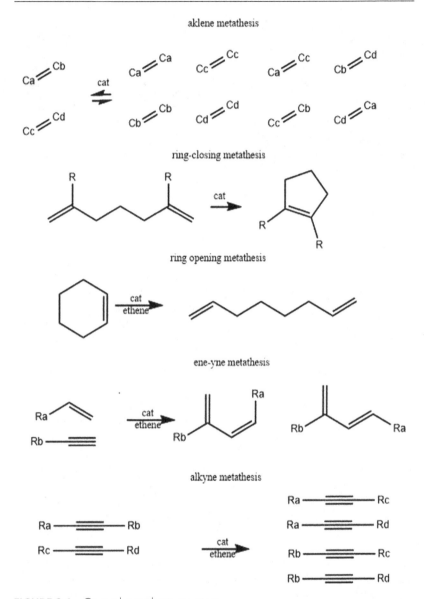

FIGURE 3.1   General metathesis reactions.

Effectively, alkene metathesis "scrambles" two alkenes. If we consider two alkenes, $C_a=C_b$ and $C_c=C_d$, the general products formed are of the type $C_a=C_c$, $C_a=C_d$, $C_b=C_c$, and $C_b=C_d$. Each of these series of products could *in principle* form two stereoisomers, leading to a total of 8 possible products though in most cases, this many is not formed. Alkyne metathesis is similar, with the important simplification that there are no stereochemical mixtures possible because of the linear geometry of the alkyne product. The end result is the same, however, from the point of view of the alkyne carbon atoms in each of the alkyne starting materials swap partners in the product.

Initial Discovery

In a way, ring-opening metathesis and the usually accompanying polymerization is the original metathesis application. As discussed in detail shortly, cyclopentene got the whole journey started and had a strong hand in Chauvin's elucidation of the mechanism. It is noteworthy that this reaction often involves the addition of ethene. This provides an additional alkene that incorporates a consistent "cap" to each carbon atom of the cyclic alkene. In truth, though any alkene would work, using an unsymmetrical alkene (needlessly) complicates the potential product mixture.

If Ring-Opening Metathesis Polymerization (ROMP) is the first, RCM is arguably the most important, if it's not the most common. It is important to notice that in this reaction, both reacting alkenes are in the same molecule, making it an intramolecular metathesis reaction. From there, the two "inside-most" alkene carbon atoms become bound by a double bond and the two "outside-most" alkene carbon atoms likewise become bound by a double bond. Many times, the alkene product generated by the combination of the two "outside" carbon atoms is volatile and leaves the system, forcing the reactions to very high yields as described by Le Châtelier's principle.

Before going into the work of Chauvin and thereafter Grubbs and Schrock, the work of Zigler and Natta (Nobel Prize in Chemistry, 1963) must be briefly mentioned as should Calderon. Ziegler and Natta earned their Nobel "for their discoveries in the field of the chemistry and technology of high polymers", polymerizing various alkenes, including cyclic ones, in a reaction now called ROMP.[1] Calderon applied the process to additional cycloalkenes and was the first to call the reaction olefin (alkene) metathesis.[2]

Chauvin's Work

Chauvin's breakthrough was built on the prior results of E. O. Fisher (Nobel Prize in Chemistry, 1973) on tungsten–carbene complexes; Natta's work on the polymerization of cyclopentene using WCl_6 and $AlEt_3$; and the work of Banks and Bailey on the formation of ethylene and but-2-ene from propene as catalyzed by $W(CO)_6$ on alumina. Chauvin and Herisson proposed the mechanism in Figure 3.2.[3] The important part of this elucidation is that it suggested that synthetically prepared metal-alkylidene complexes could be used to catalyze these metathesis reactions. To be synthetically useful, the best-case scenario is of course that such catalysts are stable enough to be commercially available; this is exactly what Schrock and Grubbs (and eventually others) went on to do with Schrock developing the first well-defined useful catalyst.

In 1964, a magical year if you ask Chauvin,[4] a number of important breakthroughs were reported. In that year, three apparently unrelated (spoiler alert … there is a relationship and Chauvin figured it out) results appeared in the literature. As Chauvin retells it, this triumvirate of discoveries were Banks's disproportionation of olefins catalyzed by molybdenum (Mo) or Tungsten (W)-based heterogenous catalysts on alumina;[5] Natta's report of the homogenously-catalyzed opening and subsequent polymerization of cyclopentene[6]; and the new metal–carbon bond demonstrating carbenes of Fischer.[7] One of Chauvin's

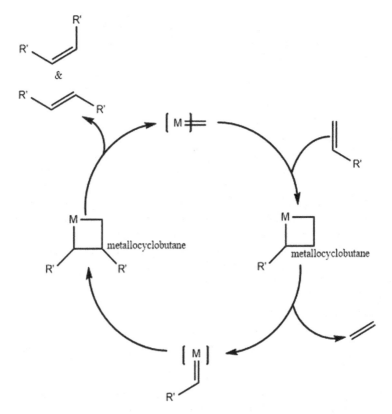

FIGURE 3.2   Mechanism of alkene metathesis.

key insights he mentions in his acceptance lecture[4] was that the disproportionation and ring opening were two parts of the same reaction. Disproportionation is a type of redox reaction where one compound is converted into two, one holding a higher oxidation state, the other a lower oxidation state than the parent compound.

Chauvin and his lab[358] ran a series of reactions that compared the telomerizaiton (a type of polymerization) and metathesis reactions using equal molar amounts of cyclopentene and pent-2-ene in the presence of one of two homogeneous catalysts: $WOCl_4/Sn(C_4H_9)_4$ or $WOCl_4/Al(C_2H_5)_2Cl$. Other than the former system giving rise to longer chain polymers, both systems gave the same mixture of products though the relative amounts of products varied slightly. Critically, the reactions gave various isomers of butene, pentene, and hexene, along with telomeriziation products with chain lengths of 9, 10, and 11; 14, 15, and 16; 19, 20, and 21; and 24, 25, and 26 carbon atoms. The first triad (C9–C11) has 2 units, the second, 3, the third 4, and the fourth, 5. Chauvin correctly reasoned that the specific sizes and repeating triads must be some sort of "residue" group (called the alcoylidine by Chauvin but since called alkylidene) must remain on the transition metal. To Chauvin (spoiler alert again ... he was right), the *most obvious* hypothesis was the formation of a metallocarbene that goes on the sequester an olefin, forming a metallacyclobutane intermediate.

Arguably, the most important part of what Chauvin elucidated is the identity of the intermediates. Though without question, a full understanding of this (and really any reaction mechanism) is very important to fully taking advantage of its synthetic utility, in this, and doubtlessly other cases, it allows for enormous synthetic utility. By knowing the identity of the intermediates, new catalyst systems could be created to leverage either thermodynamic or kinetic stabilities to make more—or less if necessary—reactive catalysts. This in principle is possible from at least two approaches. First, the intermediates themselves may be preparable and stable (enough). Second, a more stable precursor to the intermediate can be found. This ultimately is what Schrock and then Grubbs did to earn their shares of the prize. Of the two, Schrock will be covered first because his reports predate Grubbs's, even if not by very much (two years). Then, after Grubbs's contribution is covered, further developments of both and the application of metathesis to the synthesis of noteworthy synthetic targets will be covered.

Schrock's Work

For several reasons, Schrock initially turned to tantalum as the transition metal to develop a catalyst from. However, when Schrock attempted to use these species as metathesis catalysts, though they *did* form the required metallacyclobutane—a promising start—the reaction proceeded to give β-elimination products, rather than the metathesis products.[9] A breakthrough came, however, with Schrock's finding that niobium and tantalum complexes, when Cl is replaced by Ot-bu, can perform metathesis reactions with cis-pent-2-ene.[10] Next in Schrock's pursuit of an isolable catalyst was **1** (Figure 3.3), arguably the first recognizable catalyst metathesis catalyst.[11] This species successfully yielded metathesis products and importantly generated new alkylidenes. Additional pursuits of a stable but still reactive species were based on the following general understandings garnered from these studies:

1. Four coordinate species
2. Large, covalently bound ligand
3. Neopentylidene ligand
4. Avoiding alkoxides for the fourth ligand

Schrock and his lab settled on pursuing **2** (Figure 3.3). With this catalyst and related analogs, Schrock found that the activity correlated roughly with the electron withdrawing ability of the OR ligand, finding that more electron withdrawing alkoxides led to more stable intermediates that submitted to crystallization attempts and structural characterization.[12][13] They also found they were able to isolate as crystals and obtain crystal structures on intermediates that supported Chauvin's mechanism.[14] However, a complication was eventually found. In particular, the unsubstituted tungsten acyclobutane intermediate was sometimes sluggish to release the olefin. This would trap the metal and make it unavailable for additional turnover. They then made the decision to switch from tungsten to molybdenum, a fateful decision. It was with the further development of the molybdenum catalysts that Schrock et al. began to understand how to control the metathesis reaction through thoughtfully varying the

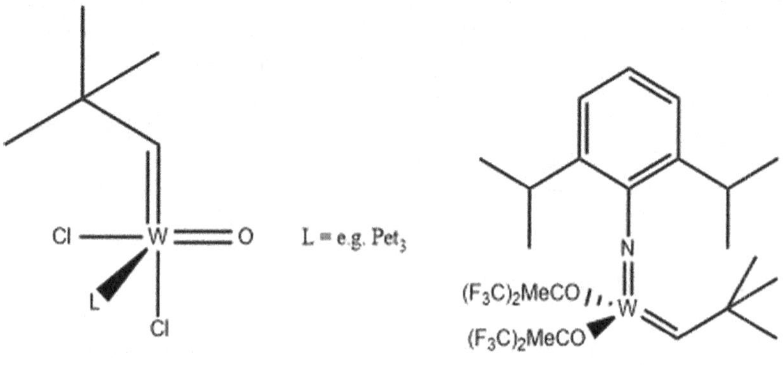

FIGURE 3.3   Schrock's initial metathesis catalysts.

size and electronic characteristics of the ligated alkoxide and imido groups. Eventually, a series of well-defined molybdenum-based metathesis catalysts were generated by Schrock et al.[15]

Grubbs's Work

We next come to the work of Grubbs. Armed with the mechanism elucidated by Chauvin, a mechanism that was further supported by work done by Grubbs[16,17] and later Katz,[18] Grubbs embarked to use rational design to optimize catalysts further. Motivated by two important shortcomings of Schrock's catalysts: (1) poor functional group tolerance and (2) the need to both prepare and handle (and of course use) the catalysts under inert atmosphere—usually in a glove box—Grubbs set out to make what can most simply be referred to as more user-friendly catalysts. These two shortcomings are related to each other though not necessarily interchangeable or concurrent in all cases. This is because the need for the inert atmosphere and/or glove box protects the reagents and reaction from a variety of atmospheric elements. In many cases, the offender is water vapor though carbon dioxide and even oxygen are common offenders as well. Functional group tolerance is important to long multistep syntheses because if reagents and/or synthetic methods that lack functional group tolerance must be used, what are called protecting groups are needed. These protecting groups do exactly what their name insinuates—they protect a group from reacting. Use of these groups adds multiple reaction steps to any multistep sequence, however, since a step must be used to both incorporate the protecting group (i.e., put it on) and to remove it to reveal the group being protected (i.e., take it off). Grubbs (rightly as far as I am concerned) reasoned that nothing less than a user-friendly reagent would fully exploit the potential power of metathesis within synthesis. The first successful ruthenium-based metathesis catalysts came out of a contemporary polymer project in Grubbs's lab. Ultimately, they found that ruthenium (II) catalysts gave outstanding results, including being functional group tolerant and functional in aqueous media. The allowance of aqueous media is extremely important as it aligns with the principles of green chemistry since it avoids the use of otherwise harmful and toxic organic solvents. This set the stage for the development of the now very well-known ruthenium (II) system of catalysts. On this journey, **3**, (a synthetic precursor for) and **4** were found to be active.[19,20] Catalyst **3** was not highly active but it was well-defined and achieved the goal of stability and functional group tolerance. Altogether, they serve as a launching point for the eventual development of **5** and **6**, now known as the first-generation Grubbs catalyst (**5**) and the second-generation Grubbs catalyst (**6**) (Figure 3.4). Grubbs attributes the remarkable tolerance to

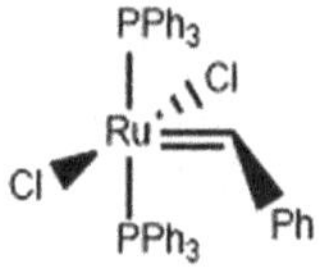

FIGURE 3.4   Initial successful Grubbs catalysts.

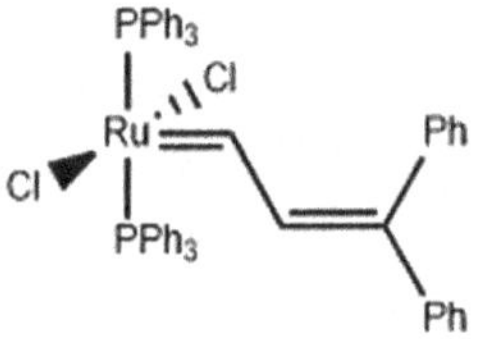

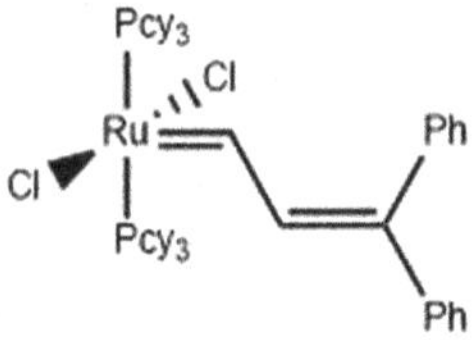

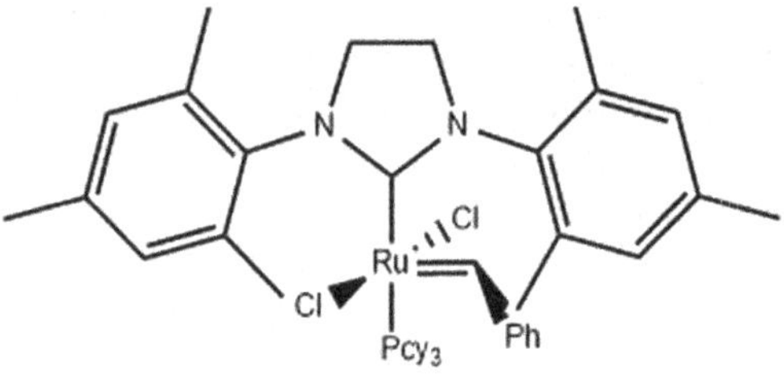

air and water to ruthenium's preference for soft Lewis acids and bases such as olefins over oxygen-based ligands such as water.

This has an additional impact in that it sort of upends much of the rules and trends learned while developing the early transition metal catalyst systems. For the less familiar reader, though ruthenium, tungsten, and molybdenum are all transition metals, tungsten and molybdenum fall into a category known as early transition metals whereas ruthenium is known as a late transition metal. This is governed by their relative positions on the periodic table and especially that the (to oversimplify) early transition metals cover half filling the five d-orbitals and the late transition metals cover filling (adding second electrons to) them. Although to best understand the specific nature of these catalytic cycles, detailed mechanistic studies were necessary, it should be recognized that nothing eventually found during these studies at all altered Chauvin's mechanism. Instead, what the detailed studies revealed was the true nature of each intermediate and other entities in this system, thereby allowing for further optimization. For example, these studies revealed that **4** and **5** were found not to be catalysts themselves but instead a catalyst precursor, generating the actual catalyst only under the reaction conditions.[21–23] These studies also rationalized the improved activity when triphenylphosphine (PPh$_3$) is replaced with tricyclohexylphosphine (Pcy$_3$)

and why the catalytic activity increases when the N-heterocyclic carbene replaces one of the Pcy_3 groups in what is commonly called the second-generation Grubbs catalyst **6**.

Modifications

Bruce Merrifield won the Nobel Prize for developing what has come to be called polymer-supported synthesis in 1984, applying this technology specifically to peptides. Since that time, the method has been extended to other types of reagents as well, among them, metathesis catalysts. The use of this approach—polymer-bound reagents—in metathesis reactions has been reviewed.[24] Briefly, this technology allows for the use of reagents that are bound to some sort of insoluble resin, often a polymer. This usually allows the reaction by-products to be removed more easily, or the easier removal of the catalyst (which often contain toxic heavy metals) because either remains bound to the insoluble polymer. This makes the reactions heterogenous. Attempts to make the use of these catalysts more environmentally friendly (which polymer-bound reagents certainly do) have been reviewed.[25] Flow reactors with homogeneous catalysts have also been reviewed[26] and are another example of highly efficient processes that reduce waste and energy use. Flow reactors can be thought of as a highly efficient continuous reaction system. Flow reactors have also been used in conjunction with soluble metal scavengers to "instantly" purify products of pharmaceutical interest.[27] The complete removal of heavy metal catalysts from products destined for the human body is extremely important from a safety perspective. Moreover, catalysts that select for the formation of the Z-alkene—which is usually not favored except in some of the smaller rings—have been reported.[28] Other stereoselective olefin metathesis work has been reviewed by Grubbs.[29] Keeping in the spirit of environmentally friendly processes, a recyclable catalyst has even been developed.[30]

New Catalysts

Schrock and Grubbs are not the only ones who have developed metathesis catalysts. A variety of others that are commercially available are shown by Strem[31] and Milipore.[32] Even a brief perusal of the line-up shows a dominance of the ruthenium-based systems. These systems are known to be more stable and easier to use so this is not surprising. A wide range of ligands, alkylidenes, and a combination of the two are used in these catalysts to tune their stability and reaction selectivity, optimized for specific purposes.

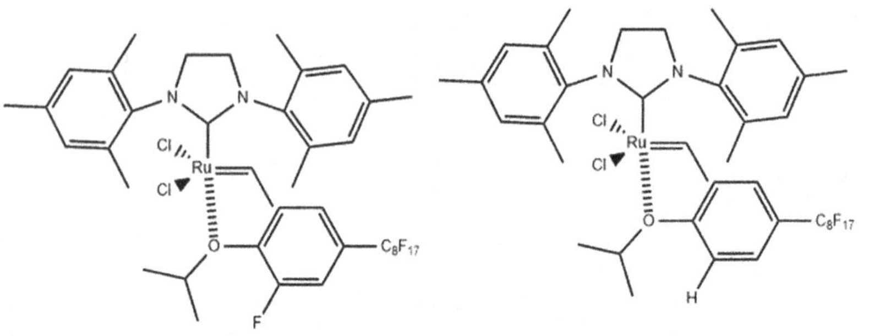

FIGURE 3.5 Matsugi's recyclable fluorescent metathesis catalysts.

Polymer-Supported Catalysts

Matsugi made a recyclable Grubbs–Hoveyda metathesis catalyst.[30] This catalyst (**8**) and a related catalyst (**7**) (Figure 3.5) further incorporates a light-activated fluorus tag. The fluorus tag not only serves an activation function but also provides a necessary handle that permits purification using fluorus solid phase extraction. Only catalyst **8** was recoverable. It should be noted that it is *not* the catalytic activity that is light-activated.

Wu and Wu[33] immobilized Grubbs-type catalysts. These novel heterogenous catalysts were immobilized on SBA-15 (Santa Barbara Amorphous), a form of mesoporous silica. In their study, Wu and Wu found that the immobilized catalysts with the largest pore size gave the best results. This was attributed to an enhanced diffusion of reactants and products.

Another study[34] uses a polyisobutylene-bound ruthenium (II) catalyst that is pyridine ligated to perform ROMP reactions. This innovative work relies on a ligand that is polymer-bound and does not dissociate from the metal. Bergbreiter[24] and Clavier and Nolan[25] discuss polymer-bound metathesis catalysts in more detail.

Continuous Flow Reactors

Continuous flow/flow reactors have also been used when employing metathesis catalysts in synthesis. For example, the work of Grela et al.[27] where continuous flow was used to generate products of pharmaceutical interest (Figure 3.6), including a large-scale preparation (>60 g) of a sildenafil (Viagra) analog, relacatib, halidor, silomat, modafinil, and pacritinib. Scale-up was made possible by running the system for 24 hours and purification was enhanced by

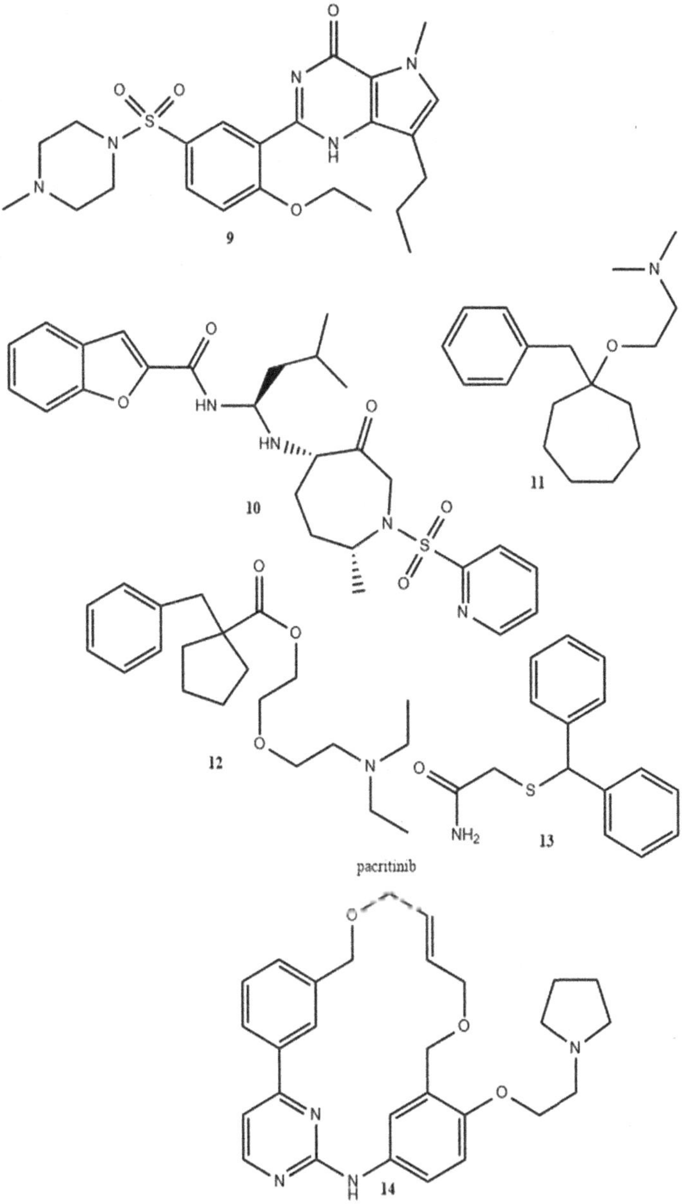

FIGURE 3.6 Pharmaceutical compounds made with alkene metathesis in continuous flow.

incorporating a metal scavenger into the system, reducing the metal contamination to <5 ppm. Reduction of heavy metal by-products or contaminants is an important step in the synthesis of pharmaceuticals and can represent a serious barrier to the large-scale production of a compound. Sildenafil (**9**) (better known as Viagra) is used to treat erectile dysfunction and pulmonary arterial hypertension.[35][36] Relacatib (**10**) potently inhibits cathepsin K[37], a key bone resorption and remodeling protease. Halidor (**11**) is an FDA-approved drug that has antispasmodic, vasodilator, and platelet aggregation inhibitor properties,[38] making it useful in circulatory disorders. Silomat (**12**) is used as a treatment for asthma.[39] Modafinil (**13**) is used to treat sleepiness arising from a variety of causes.[40] Finally, pacritinib (**14**), a kinase inhibitor, is used to treat myelofibrosis and lymphoma.[41]

APPLICATIONS

Metathesis reactions, especially RCM has been used to prepare many important synthetic targets. Even to say that these reactions opened up syntheses previously impossible or at best requiring far more steps or less efficient ones would still be an understatement. Alas, there is only but so much space that can be devoted to one class of reactions in a short volume such as this. For additional information, Vanderwal and Atwood have published a review of advances in alkene metathesis for natural product synthesis.[42] Meanwhile, Enders et al. penned a review on synthetic strategies to access specifically seven-membered carbocycles,[43] an area that metathesis figures prominently, even nearly exclusively. A review of the patent literature (which is sometimes difficult to navigate for a novice) regarding olefin metathesis in drug discovery and development has also been written.[44] It's impossible to cover in a small volume like this even a representative compilation. Here, the application of metathesis to the synthesis of materials, natural products, and pharmaceuticals will be covered briefly.

Materials

In addition to a plethora of natural products and pharmaceutical compounds, metathesis reactions have been used to prepare a host of materials, particularly materials with electronic properties as well. One example are the helicenes. Helicenes (e.g., **15**), which are best known for their electronic and optical properties have also been prepared using RCM (Figure 3.7).[45] In an unrelated

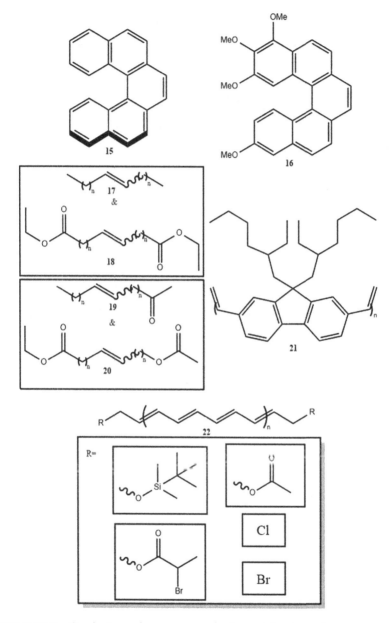

FIGURE 3.7 Conducting polymers prepared using metathesis reactions.

work, substituted helicenes like the 5-helistatins analog (**16**) (Figure 3.7) have recently been shown to have tubulin-binding antimitotic activity.[46]

Cross metathesis has been used in the metathesis of renewable materials as an alternative feedstock for critical building blocks for chemical manufacturing.[47–57] One example of use to generate an alternative feedstock for critical building blocks is the work of Grisi et al.[58] in the cross metathesis of unsaturated fatty acids (Figure 3.7). These processes led to α-, ω-difunctional compounds (**17–20**) that can serve as purifications for polymerization reactions. Many of the unsaturated fatty acids used for such a process are harvestable from natural sources that are farmed and grown annually. Such α-, ω-difunctional compounds have functionality at the two ends of a molecule or, more biblically, at the beginning (alpha) and end (omega) of the molecule.

Alkene metathesis has also been used in the synthesis of conducting polymers; Kumar et al.[59] have reviewed this topic. Conducting polymers are semiconductors whose use and preparation has been reviewed extensively including but not limited to Kumar (as above), Rout,[60] and others. They have been used in a wide array of electronic fields such as sensors, organic electronics (e.g., photovoltaics and light emitting diodes), thin-film transistor applications, and many others as cited in the reviews above and elsewhere. They've even been used as chemical reagents in reactions of alcohols such as Ritter reactions between alcohols and acetonitrile,[61] Friedel–Crafts reactions between of alcohols with benzene and toluene,[62] homoetherification of alcohols,[63] and cyclodehydration of sorbitol.[64]

Acyclic diene metathesis (ADMET) has been extensively used in the synthesis of conducting polymers; synthesizing meaning here the polymerization step. ROMP has also been used to polymerize monomers into conducting polymers. One example (Figure 3.7) of the use of ADMET is the synthesis of poly-(fluorenylenevinylen)s (**21**),[65] conducting polymers explored for light emitting diode applications.[66]

One example of ROMP[67] being used in the synthesis of conducting polymers is in the preparation of telechelic polyenes from cyclooctatetraene and appropriate end-capping reagents provided by a symmetrical alkene or polymer with a vinylic end cap; the former is shown in Figure 3.7. Intrinsically coupled polymers like these (**22**) are used for a variety of purposes such as light emitting diodes and charge storage devices.

Natural Product Synthesis

Blechert et al. used a tandem ene-yne-ene-RCM sequence (Figure 3.8) to complete the total synthesis of ent-(enantiomer of) lepadin F (**23**) and G (**24**).[68] The lepadins were first identified as secondary metabolites of the North Sea

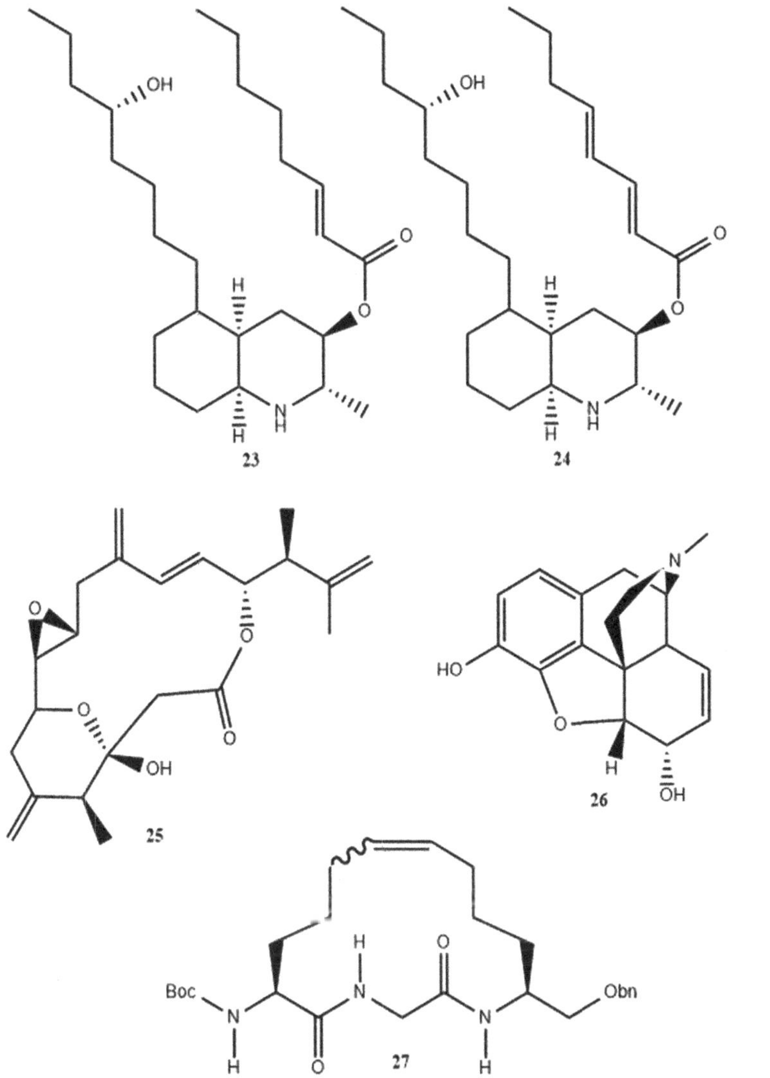

FIGURE 3.8 Natural products synthesized using metathesis reactions.

tunicate Clavelina lepadiformis in the early 1990s and were subsequently found to have activity against human cancer cell lines.[69,70]

Another example is the use of ene-yne metathesis by Jecs and Diver[71] (Figure 3.8) to achieve the total synthesis of amphidinolide P (**25**). In their synthesis, they used two metathesis methods, ene-yne metathesis and RCM using two different approaches to the target. Amphidinolide P is a natural product

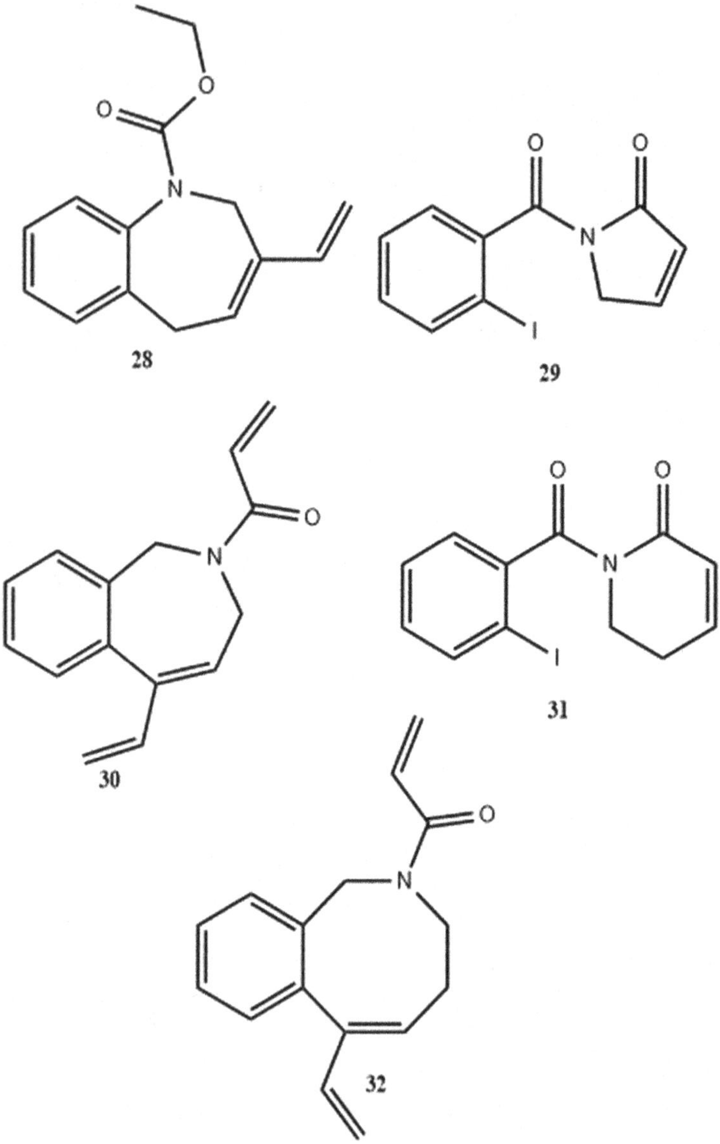

FIGURE 3.9   Benzazepines, lactams, and benzazocines made by metathesis.

found in the marine dinoflagellate *Amphidinium sp.* This compound has been shown to have mild cytotoxicity against a form of leukemia and a form of epidermoid carcinoma.[72-76]

Morphine (**26**) (Figure 3.8) has been synthesized by Smith et al.[77] (and many others) and is on the World Health Organization's list of essential medicines. A key series of steps employs a cascade ene-yne-ene-RCM sequence. Using the second-generation Grubbs catalyst, an ene-yne metathesis first occurs to give the B ring and this is followed by the alkene RCM to give the C ring. Subsequent deprotection and other chemical elaboration steps gave morphine. Morphine and its derivatives (like heroin) are a much maligned class of compounds, presently. It is fair to wonder why we would need *another* synthesis of it. To date, the only viable commercial (and illicit, it turns out) source of this essential medicine is isolation from its natural source. Although there is no sign right now that the opium poppy is threatened by climate change, if it is ever to be, the primary source of an essential medicine would be at risk. Also, geopolitical conflicts can disrupt these sorts of global supplies when the conflict engulfs regions that are the primary growing source of these plants. Thus a synthetic route that would compete with the natural source would be very desirable.

Demonstrating versatility in the range of targets possible, quite early in the use of RCM, it was shown that cyclic tripeptides (**27**) can be made (Figure 3.8), in this case furnishing a compound that showed micro to nanomolar inhibition of *Rhizopus chinensis* pepsin.[78] These compound represented a new class (at the time) of simplified aspartic protease inhibitors.

Benedetti et al.[79] used metathesis reactions (Figure 3.9) to gain easy access heterocyclic compounds. Using this model system, 1-(**28**) and 2-(**29**) benzazepine scaffolds can be accessed. Other starting materials could be used to make lactams five (γ) (**30**) or six (δ) (**31**) atoms and a benzazocine derivative (**32**). Each of these motifs are common moieties in pharmaceutical compounds. Synthetic sequences that allow for easy access to a wide range of differently functionalized derivatives are highly valuable for this reason.

SUMMARY

The synthetic power of the chemical transformations these catalysts make possible is truly immeasurable. Part of why this is the case is that prior to the development of the RCM method, rings of certain sizes were difficult to make; these ring sizes are routine for metathesis. The synthesis of countless natural products and other targets have been made shorter because of the synthetic methods enabled by the catalysts made by Schrock and Grubbs. None of their work would be possible (or would have taken far longer) if not for the work of Chauvin.

REFERENCES

(1)   *Makromolecular Chemistry* **1966**, 87.
(2)   *Accounts of Chemical Research* **1972**, 127.
(3)   *Makromolecular Chemistry* **1971**, 161.
(4)   Olefin Metathesis: The Early Days. www.nobelprize.org/uploads/2018/06/chauvin-lecture.pdf, last checked 8/10/24.
(5)   *Comptes rendus de l'Académie des Sciences Paris* **1973**, 169.
(6)   *Makromolecular Chemistry* **1976**, 2637.
(7)   *Journal of the Chemical Society Chemical Communications* **1992**, 462.
(8)   *Comptes Rendus Hebdomadaires des Seances de Serie C* **1973**, 169.
(9)   *Journal of the American Chemical Society* **1979**, 4558.
(10) *Journal of Molecular Catalysis* **1980**, 73.
(11) *Journal of the American Chemical Society* **1980**, 4515.
(12) *Organometallics* **1984**, 1563.
(13) *Organometallics* **1984**, 1554.
(14) *Progress in Inorganic Chemistry* **1991**, 1.
(15) *Journal of the American Chemical Society* **1990**, 3875.
(16) *Journal of the American Chemical Society* **1976**, 3478.
(17) *Journal of the American Chemical Society* **1975**, 3265.
(18) *Journal of the American Chemical Society* **1976**, 2519.
(19) *Journal of the American Chemical Society* **1992**, 3974.
(20) *Journal of the American Chemical Society* **1993**, 9858.
(21) *Journal of the American Chemical Society* **1997**, 3887.
(22) *Journal of the American Chemical Society* **2001**, 749.
(23) *Journal of the American Chemical Society* **2001**, 6543.
(24) *Polymers* **2016**, 140.
(25) *Angewadte Chemie International Edition* **2007**, 6786.
(26) *Green Chemistry* **2012**, 2012.
(27) *ACS Sustainable Chemistry & Engineering* **2021**, 16450.
(28) *Nature* **2011**, 461.
(29) *Catalysts* **2017**, 87.
(30) *Journal of Organic Chemistry* **2010**, 7905.
(31) Metathesis Catalysts. https://media.abcr.com/pdf/strem-metathesis-catalysts.pdf, last checked 8/11/24.
(32) Metathesis. www.sigmaaldrich.com/US/en/technical-documents/technical-article/chemistry-and-synthesis/metathesis/metathesis, last checked 3/25/24.
(33) *Journal of Molecular Catalysis A: Chemical* **2013**, 45.
(34) *ACS Omega* **2016**, 714.

(35) *European Respiratory Journal* **2006**, 563.
(36) *Sexual Medicine Reviews* **2019**, 115.
(37) *Journal of Medicinal Chemistry* **2006**, 1597.
(38) *Thrombosis Research* **1974**, 741.
(39) *Current Therapeutic Research, Clinical and Experimental* **1964**, 14.
(40) *Drugs* **2008**, 1803.
(41) *Lancet Haemetology* **2017**, e225.
(42) *Aldrichimica Acta* **2017**, 17.
(43) *Synthesis* **2013**, 845.
(44) *Organic Processes Research and Development* **2017**, 1938.
(45) *Angewadte Chemie International Edition* **2006**, 2923.
(46) *JACS Au* **2022**, 2561.
(47) *Applied Catalysis A* **2009**, 32.
(48) *Bielstein Journal of Organic Chemistry* **2011**, 1.
(49) *ChemSusChem* **2017**, 470.
(50) *Chemistry A European Journal* **2016**, 12226.
(51) *European Journal of Lipid Science and Technology* **2011**, 39.
(52) *Green Chemistry* **2007**, 1356.
(53) *Green Chemistry* **2011**, 2911.
(54) *Green Chemistry* **2014**, 1579.
(55) *Monatshefte für Chemie* **2012**, 669.
(56) *RSC Advances* **2016**, 100925.
(57) *RSC Advances* **2014**, 55622.
(58) *Catalysts* **2020**, 904.
(59) *International Advanced Research Journal in Science, Engineering, and Technology* **2015**, 110.
(60) *RSC Advances* **2021**, 5659.
(61) *Synthetic Communications* **2013**, 3224.
(62) *Polymers* **2007**, 4328.
(63) *Journal of Polymer Science A: Polymer Chemistry* **2007**, 2328.
(64) *Synthetic Metals* **2010**, 2284
(65) *Organometallics* **2008**, 1479.
(66) *Macromolecules* **2002**, 7532.
(67) *Journal of the American Chemical Society* **2003**, 8515.
(68) *Journal of Organic Chemistry* **2008**, 3088.
(69) *Tetrahedron* **1991**, 8729.
(70) *Tetrahedron Letters* **1995**, 6189.
(71) *Organic Letters* **2015**, 3510.
(72) *Journal of Organic Chemistry* **1995**, 6062.
(73) *Natural Products Reports* **2004**, 77.
(74) *Journal of Natural Products* **2007**, 451.
(75) *Journal of Antibiotics* **2008**, 271.
(76) *Israeli Journal of Chemistry* **2011**, 329.
(77) *Angewadte Chemie International Edition* **2016**, 14518.
(78) *Bioorganic Medicinal Chemistry Letters* **1998**, 357.
(79) *Synthesis* **2012**, 3523.

2010
Heck, Negishi, and Suzuki

4

The 2010 Nobel Prize in Chemistry was awarded for palladium-catalyzed cross-couplings in organic synthesis to three people: Richard Heck, Ei-ichi Negishi, and Akira Suzuki. These reactions, and other palladium-catalyzed coupling reactions not recognized in this award, are among the most important in organic synthesis. This is because they form carbon–carbon bonds and there is no apparent limit to the carbon–carbon single bond that can be made using these combined methods. These reactions are called cross-coupling reactions because they couple together two different pieces. The types of single bonds that can be made via these processes include bonds between sp^3 and sp^3 hybridized carbons, sp^3 and sp^2 hybridized carbons, and sp^2 and sp^2 hybridized carbons. Even sp hybridized carbons participate in these coupling reactions. Before covering their synthetic utility, however, it is worth mentioning why palladium is so good at catalyzing these kinds of reactions.

In truth, palladium is not alone in its ability to catalyze reactions, even the formation of carbon–carbon bonds is not something that palladium has exclusive ownership of. An earlier chapter in this very volume in fact covered ruthenium and molybdenum catalyzing the metathesis of olefins. Palladium-catalyzed coupling reactions have much more versatility and importantly form carbon–carbon *single* bonds. Something about palladium must therefore be special.

Ions of palladium (II) are special electronically and structurally in that they adopt a d^8 electron configuration and a square planar geometry. This places two coordination sites in the axial position, a wide open position for a ligand or other reagent to approach unencumbered. Palladium orbital energies are also very well-tuned to coordinate with olefins. It turns out, platinum and nickel likewise have similarly tuned orbital energies but platinum suffers from poor kinetics (reaction rates) and to date, palladium has outpaced nickel in terms of how well reactions work though there are case-by-case exceptions

DOI: 10.1201/9781003006879-4

to that generality. Palladium (as a neutral element) is also *unique* electronically. Despite the overgeneralizations about electron configuration sometimes encountered at the introductory level, the transition metals do some unexpected, even weird things as they fill their electron orbitals during the march through the periodic table. Even as transition metals go, palladium is somewhat of an oddball since its ground state electron configuration has all five d orbitals filled ($4d^{10}$) and an *empty* 5s orbital. The electronic oddities of palladium don't end there—palladium's *lowest* d to p transition is atypically large. All of this combines, along with the geometries adopted by the complexes of cationic palladium species to permit very interesting chemistry.

Palladium was also part of an earlier Nobel Prize for Paul Sabatier in 1912, covered in Volume 1 of this series. While Sabatier's prize was for the hydrogenation of olefins using finely divided palladium, this award is given for the *creation* of carbon–carbon single bonds. It is also noteworthy that although there are other Nobel-winning transformations in the realm of organic synthesis, these methods are one of a precious few (five to be precise) Nobel-winning transformations that generate carbon–carbon bonds—the Grignard reaction (1912, volume 1), the Diels–Alder reaction (1950, volume 1), the Wittig reaction (1979, volume 2), and olefin metathesis (2005, this volume).

NOBEL-WINNING PALLADIUM CROSS-COUPLING REACTIONS

In general, one case of the palladium cross-coupling reaction couples (joins) together two carbon atoms where one of the carbon atoms has a metal (typically Zn, Sn, or B (a metalloid)) and the other carbon atom has a leaving group, typically a halogen or sulfonate ester (Figure 4.1A and B). The Heck reaction (Figure 4.1A) can be summarized as follows. Though commonly used to prepare conjugated systems (where an alkene is conjugated with another alkene or aryl system), it can be used to alkylate alkenes as well at the less substituted position of the alkene. The stereochemical outcome of the reaction is *always* that the added group is in the same stereochemical position as the hydrogen atom it replaces. If an alkyl group is to be used as the halide, there *must not* be a hydrogen atom in the β position. In general, the reaction proceeds at a slower rate with chlorides, compared to bromides.

Meanwhile, in summary, in the Negishi cross-coupling (Figure 4.1A) the organozinc reagent typically requires an additional step to prepare via reaction of the halide with activated zinc or by the reaction of the corresponding Grignard reagent with a zinc halide salt. While the joining of two sp^2 hybridized carbon

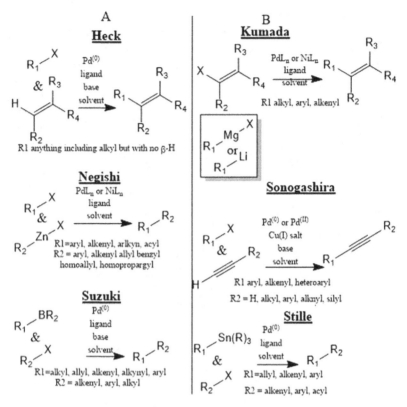

FIGURE 4.1 (A) Nobel Prize-winning Pd-catalyzed reactions; (B) other Pd-catalyzed coupling reactions.

atoms is the most common use, the coupling of sp^2-sp and sp^3-sp^2 hybridized carbons are both also possible.

Finally the Suzuki reaction (Figure 4.1A) can be described as being able to generate a wide range of carbon–carbon single bonds across any carbon hybridization combinations. Although like the Negishi coupling (with the requisite formation of the Zn reagent) this reaction too requires the preparation of the boronyl starting material, many such reagents that are common structural moieties are now commercially available; this is likely due to the importance of this very reaction. Other noteworthy aspects of this reaction include that the reagents are generally safer (to humans and the environment) and reaction by-products are usually easy to remove. The former is an obvious advantage and the latter is extremely important when preparing targets on a large scale since it makes purification easier and cheaper.

OTHER PALLADIUM-CATALYZED CROSS-COUPLING REACTIONS THAT MAKE CARBON–CARBON BONDS

Although only Heck, Negishi, and Suzuki are recognized with a Nobel Prize, there are other palladium-catalyzed carbon–carbon bond forming reactions worth mentioning (Figure 4.1B). It is unlikely that any of these will likewise earn a Nobel Prize for those who discovered them. This is because the level of improvement necessary to justify a second Nobel for essentially the same work simply does not exist. If attempts to broadly apply any one of them to standard modes of synthesis or to apply new technologies like artificial intelligence (AI) are successful, this may change, but for variations of the reaction, I think the Nobel book is closed for palladium-catalyzed cross-coupling reactions.

Other reactions include the Kumada, Sonogashira, and Stille coupling reactions. In the Stille reaction, catalyzed by palladium, an organotin compound is coupled with a halide. Both partners typically have an sp^2 or sp hybridized carbon atom as the reacting partner and a useful variation can insert a carbonyl between the two reacting partners. Tin, however, is toxic so it's widespread use in the synthesis of compounds for human consumption is limited. The Sonogashira reaction couples an sp or sp^2 carbon centered halide or triflate with a terminal alkyne, leading to the formation of a new single bond between sp and sp or sp^2 and sp hybridized carbon atoms. The Kumada reaction is similar to the Heck in that an olefin is one of the reacting partners but differs in the use of a Grignard or organolithium reagent. Also, the leaving group is on the olefin where in the Heck reaction, the olefin is the nucleophile. In a way, this reaction can be considered an umpolung (reverse polarity) Heck reaction. Finally, the Buckwald–Hartwig amination (not shown) deserves at least brief mention since it is a way to reliably make aryl amines and aryl halides under palladium catalysis. This reaction is important to mention at least briefly because aryl amines are common structural features in pharmaceutical products and natural products alike.

Mechanistically, in general, the two carbon atoms destined to be coupled initially form some manner of carbon–metal bond with the palladium. This means that in the case of the moiety already bound to a metal, the palladium displaces the other metal and in the case of the moiety that contains the leaving group, displaces said leaving group instead. The end result is what is most important—both carbon atoms are directly bound to the central palladium metal. The close proximity made possible by this is what leads to their coupling as they dissociate from the metal in a process that will be made more clear when discussing the mechanisms later. In another case, the electrophilic partner is still a halogen or sulfonate ester while the nucleophile is an olefin (an alkene). In general, each component is coordinated with the palladium in turn and the new bond forms between the two carbon atoms, resulting in the dissociation of each from the palladium.

Of the reactions covered in this chapter, the work of Heck will be covered first, if for no other reason, it came first. Negishi's work will be covered after that and then finishing with the work of Suzuki. This is the order in which the reactions were discovered and it is simply coincidental that it also represents the alphabetic ordering of the reactions.

THE HECK REACTION

Heck's work in the area of transition metal chemistry (and eventually palladium) began in 1956 when Dr. David Breslow (Heck's supervisor at the time) suggested he begin work on transition metals. Heck's first foray—investigating the hydroformylation process—resulted in what is regarded as the first elucidation of a transition metal-mediated reaction mechanism.[1] Heck's move to palladium was instigated by another colleague, Pat Henry, who was studying the mechanism of the Wacker Process: a process that oxidizes ethylene to acetaldehyde. Heck set out to investigate what would happen if the organopalladium species lacked a hydrogen atom in the beta position, relative to the metal in the presence of another molecule. Heck observed the formation of stilbene and styrene upon mixing phenyl mercuric acetate with tetrachloropalladate under an ethylene atmosphere.[2]

Heck's pioneering work was first reported in 1968. In this seminal work, he showed that methyl[3] and phenyl[4] halides can be coupled with olefins. This initial work was non-catalytic, generating palladium metal during the reaction. That same year, however, Heck reported that with the addition of $CuCl_2$ in the reaction, the palladium metal is reoxidized, allowing for the regeneration of the palladium species and making the process catalytic.[2] Heck wasn't done here. He also elucidated the mechanism of the reaction and accounted for the stereochemical outcome—that the trans, *less hindered*, alkene is formed—of the reaction.[5] He later (1972) published a modified version of the reaction, now the standard protocol.[6] In this version, the RPdX species is generated from reacting the organic halide (RX) with neutral palladium in what is referred to as an oxidative addition, so called because during the addition, the metal is oxidized.

The Heck reaction is noteworthy for several reasons. The conditions are very mild; neither high nor cryogenic temperatures are needed, harsh acids and bases are not employed and they can be carried out in open vessels.

The Heck reaction mechanism (Figure 4.2) begins with what is called an oxidative addition as the palladium is oxidized to the 2+ oxidation state upon reaction with the ligated palladium (1) with the organohalide (2), inserting itself between the carbon and the X and losing two electrons. The olefin (3) then participates in what is referred to as a *syn* migratory insertion, so called because the palladium and the R group are added to the olefin on the same face of the former alkene. This is followed by a rotation of the now carbon–carbon single bond to place the β-hydrogen atom and palladium *syn* to one another in the intermediate. This elimination (re-)creates the olefin and released the palladium. When another carbon atom α to the palladium (next door to the palladium atom's carbon) has a hydrogen atom (a so-called β-hydrogen), a similar carbon–carbon bond rotation can place this alternative hydrogen atom *syn* to the palladium, resulting in the possibility that the olefin forms in this direction as well, giving a mixture of products. The addition of a base allows for the HPdX by-product to regenerate the catalytic Pd species by reducing the palladium 2+, starting the process over.

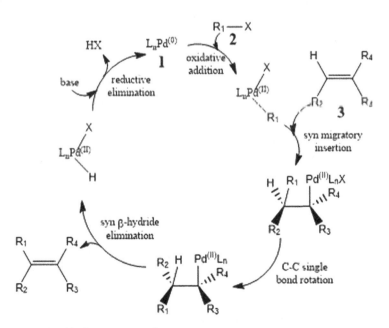

FIGURE 4.2 Heck reaction mechanism.

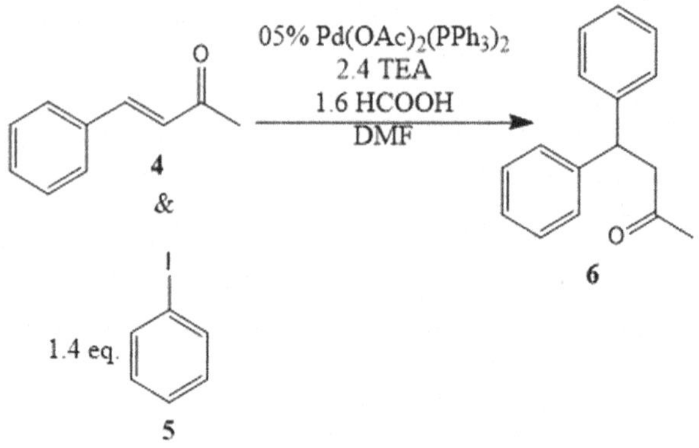

FIGURE 4.3   Reductive Heck reaction.

REDUCTIVE HECK REACTION VARIATION

Cachi and Arcadi[7] first reported conditions that are a variation of the Heck reaction and have now come to be called a reductive Heck reaction (Figure 4.3). In their work, benzalacetone (**4**) was reacted with iodobenzene (**5**) with a palladium catalyst, a base, and a reducing agent to give what appointed to the conjugate addition-type product (**6**). So popular has this method become that its use in the synthesis of natural products has been reviewed.[8]

NEGISHI CROSS-COUPLING

The next Nobel-winning palladium cross-coupling work was developed by Negishi. In 1976, Negishi began studies that further developed the earlier work of Kharasch,[9] who in 1941 used cobalt, nickel, and iron salts (later found to be due to homocoupling of the Grignard reagent starting material); the work of Gilman[10] using copper salts to couple MeMgBr and MeI; Kochi's iron catalyzed work,[11] Corriu's[12] and Kumada's[13] nickel work; and finally Murahashi's extension of the work to Pd.[14] In 1976, Negishi began studies using organozinc and organoaluminum compounds in palladium-catalyzed reactions.[15,16] As part of this survey, in 1977, Negishi turned to organozinc compounds.[17,18] These reagents proved higher-yielding, very mild, and more selective, a triumvirate of "betterness." One significant advantage permitted by the organozinc reagents over harsher reagents like Grignard reagents and organolithium reagents is that they (the zincates) are more tolerant of a wider range of functional groups.

Negishi also found during these studies that alkynyl boron compounds also coupled with organohalides in the presence of palladium catalysts but did not pursue them further. Suzuki, on the other hand, *did* pursue them further and so the method that employs these reagents now bears the latter's name and is the subject of the next section.

Negishi's first success came without using any palladium, instead using alkenyl boronates to generate conjugated E and Z dienes in very impressive yields.[19-21] Initial attempts to explore the use of cuprous halides to further this synthetic methodology failed due to impure cuprous halides, unknown as the cause at the time. Inspired by contemporary work by Tamao using nickel-catalyzed Grignard cross-couplings,[132223] they explored the coupling of Grignard reagents with alkenyl borons and borates, only to fail again. When aryl bromides and iodides replaced the boron reagents, smooth coupling of alkenylalanes in the presence of $Ni(PPh_3)_4$ and $Pd(PPh_3)_4$ catalysts were observed.[24] In these studies, the palladium-catalyzed reactions did not fair markedly better than their nickel counterparts however, upon extending these studies, the palladium-catalyzed reactions gave superior retention of the original alkenyl geometry, even if the nickel variant only gave up to 10% of the undesired product.[16] Though this admittedly small amount may seem insignificant to the novice, if the reaction in question is on a multi-kilogram scale batch of commercial products, a large and unacceptable expense may be suffered by the manufacturer. This work provided the first case of cross-coupling reactions with non-Grignard reagents catalyzed by nickel and palladium. Negishi et al. then explored a range of metal partners, reasoning that the organometallic reagent served the important purpose of delivering the R group to the palladium species. Although they explored a range of metals and metalloids (Li, Mg, Zn, Hg, B, Al, Sn, Zr), zinc, boron, and tin were found to work best.[2526] This study supported Negishi's earlier finding of the effectiveness of zinc compounds in these coupling reactions.[17] Meanwhile, Sn-catalyzed reactions later further developed by Stille but also reported by Kojugi[27] and Suzuki who further developed the boron version described in this chapter.

The Negishi cross-coupling reaction mechanism (Figure 4.4) starts as the Heck does with an oxidative addition to ligated palladium (**1**) by an organohalide (**2**). The organozinc reagent (**7**) then delivers the other moiety, resulting in both carbon groups being on the palladium metal. A reductive elimination then occurs, forging the new carbon–carbon bond and reducing the oxidized palladium species to regenerate the neutral metal catalyst, unlike the Heck reaction which needs another additive to do this catalytic turnover step. Also, unlike the Heck reaction, this reaction does not result in an alkene (re-) forming in multiple directions.

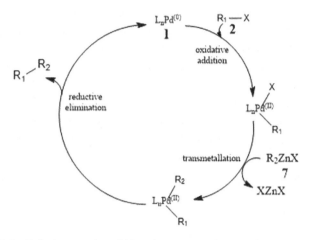

FIGURE 4.4   Palladium-catalyzed Negishi cross-coupling mechanism.

SUZUKI CROSS-COUPLING

The Suzuki reaction is also called the Suzuki–Miyaura cross-coupling reaction but will be referred to hear as the Suzuki cross-coupling. In 1979, Suzuki and his lab reported the coupling of organoboron compounds with aryl halides with later work extending the scope to alkyl groups as well.[28,29] Aklyl/aryl boronic acids[30] and their ester counterparts, it was later found, also serve as competent coupling partners to generate biaryls.[31] This, for reasons discussed more later, further extends the utility of the Suzuki reaction.

Of the palladium cross-coupling reactions, discussed here, the conditions now referred to as the Suzuki coupling/reaction are perhaps the most versatile. This versatility is due to the reality that any organoborane reagent will react with anything with a suitable leaving group, commonly a halide. Thus as long as one of the reacting carbon atoms can be bound to boron and the other to a halide—and this is virtually if not literally always possible—the two carbon atoms can be joined using these conditions. Effectively, this method can be used to create *any* carbon–carbon single bond.

Recall that Negishi had initially explored boron-mediated coupling reactions, though the reactions did not proceed smoothly with palladium under his conditions. It was reasoned that for trialkyl boranes, the carbanion character of the would-be nucleophilic carbon was not strong enough. However, a study by Gropen and Haaland found that for tetraalkyl boranes, such as tetramethylborane, the methyl group was 5.5x more electronegative.[32] As such an increase was also expected in the presence of a base, Suzuki and co-workers

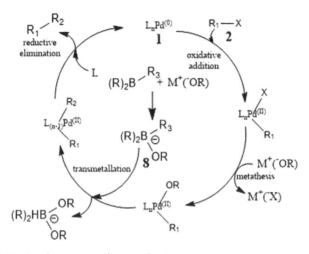

FIGURE 4.5 Suzuki cross-coupling mechanism.

explored the reaction in the presence of a base and observed excellent yields of the cross-coupling product.[28][29]

Like both the Heck reaction and Negishi cross-coupling, the Suzuki cross-coupling mechanism (Figure 4.5) starts with an oxidative addition to ligated palladium (**1**) by an organohalide (**2**) where each the R group and the X are added to the palladium. A metathesis (not the kind discussed earlier in this volume) then occurs where the leaving group attached to the palladium is replaced with the base. A transmetallation reaction then occurs whereby the other organic group on the activated boronyl species (**8**) is delivered to the palladium; that the palladium replaces the boron from the carbon atom's point of view is why this is a transmetallation. Reductive elimination with concurrent readdition of the ligand releases the product and regenerates the catalytic species.

These reactions also permit the coupling of heteroaromatic molecules. Such molecules are aromatic, like benzene, but contain a heteroatom (O, N, S) within the ring. Access, particularly efficient and high-yielding access to these compounds, is very important given their common presence in natural products and pharmaceuticals.

MODERN DEVELOPMENTS

The developments described here are by no means the only modern developments that extend this important work. Nevertheless, quite arbitrarily,

I selected those most interesting to me or that I've had a personal encounter with. Work from the Leadbeater lab where microwave-assisted heating permits extremely low levels of catalyst to be used will be described. After that, processes that permit these important coupling reactions to proceed with enantioselectivity will be presented, followed by developments that allow for the recycling and reuse of catalysts. This section on modern developments will conclude with a brief description of the exciting work coming from the Burke lab that employs AI and machine learning, along with pursuits of automated iterative carbon–carbon bond formations that will be detailed. Modifications and other improvements, though, continue all the time. Recently, Andrade and Martins[33] reviewed the implementation of methodologies they deemed unconventional in carbon–carbon bond forming cross-coupling reactions reported in the literature over an approximately five-year period starting in 2015. Methods included in their review are: microwave heating and sonochemistry as activation methods and solvent-free, aqueous media, poly ethylene glycol, ionic liquids, and deep eutectic solvents as alternate media for these important reactions. There have also been attempts to replace palladium with other metals.

Leadbeater and the Use of Microwave Heating and Flow Reactors

The story of the ultra-low palladium level Suzuki-type reactions starts in 2003[34–36] when the Leadbeater lab reported Suzuki-type couplings without adding palladium catalysts under microwave heating conditions. Around this time, Professor Leadbeater was interviewing for a position at the University of Connecticut where I was pursuing and ultimately earned my PhD; Professor Leadbeater remains at UCONN today. I remember well part of the title of his public talk being OrgaNOmetallic chemistry. Subsequent work, however, revealed that ultra-low levels of palladium were present in their sodium carbonate used as the base in their biaryl-forming protocol.[37] This is a known contamination of commercial sodium carbonate. It is noteworthy, however, that the use of microwave heating gave superior yields than conventional heating even when using ultra-low levels of added catalyst later. The reaction protocol has the added advantage of using water as the primary solvent. They also used microwave heating to generate nonsymmetrically substituted stilbenes in a one-pot two-step Heck strategy.[38] In this protocol, an aryl halide is first coupled with ethene and after the first reaction is complete, the second aryl halide is added to the pot and heating is resumed. The sealed tube of the microwave-assisted organic synthesis allowed for the loading of the system with ethene during step one, taking advantage of Le Châtelier's principle to consume all the initial aryl halide. After step one was over, the system was safely vented to remove all the excess ethene, with the addition of the second aryl halide following. After heating, the unsymmetrical stilbenes were found in very good yield in most cases. The Leadbeater lab also has pursued Suzuki couplings with microwave heating in a flow system for large batch synthesis.[39] When using a 1:1 ethanol:water mixture as the solvent—done in order to maximize the solubility of the organic reagents—the solubility of mineral bases such as sodium carbonate is compromised. By switching to organic bases such as TMF (1,1,3,3-tetramethyl guanidine), DBU (1,8-diazobicyclo[5.4.0]undec-7-ene), or DABCO (1,4-diazobicyclo[2.2.2]octane), good yields were seen; DABCO having the lowest yields. These bases all remained soluble in the solvent system, avoiding any issues that stopped or inhibited the flow of reaction mixtures due to the precipitation of insoluble salts. Finally, the Leadbeater lab used microwave heating to carry out solvent-free, open-vessel microwave-promoted Heck reactions.[40]

Enantioselective Palladium-Catalyzed Cross–Cross-Coupling Reactions

As discussed in an earlier chapter in this volume and elsewhere in this series, stereoselectivity in reactions is ardently pursued. Considering the importance the reactions in this chapter hold—that they facilitate the formation of a carbon–carbon bond—methods that permit their formation stereoselectively carry added value since they can be used to stereoselectively make carbon–carbon bonds even in the absence of a functional group in the product at the chiral center made. The use of enantioselective palladium-catalyzed cross-coupling reactions has been reviewed, including but not limited to Biscoe et al.,[41] Gury and Kelly,[42] Shibasaki et al.,[43] and Reisman et al.[44] For example, Phipps et al.[45] who developed an enantioselective Suzuki–Miyaura coupling to generate axially chiral phenols (9). In their study, a sulfonated chiral ligand sSphos was employed (Figure 4.6). After optimization, yields were in the 70% range with %ee values over 90%.

Cormona et al.,[46] on the other hand, used a dynamic kinetic asymmetric Heck reaction to simultaneously generate central and axial chirality (Figure 4.6). In this reaction, the isoquinoline couples to 2,3-dihydrofuran to give a heteroaryl 3,4-dihydrofuran derivative (10). The central chirality (on a stereocenter) is the chiral center on the dihydrofuran ring in the product while the axial chirality is found in the isoquinoline system of the product. To generate this chirality, in situ formed [Pd^0/DM-BINAP] or [Pd^0/Josiphos ligand]

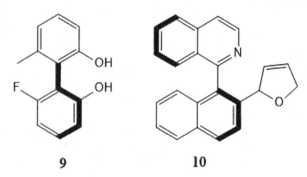

9   **10**

FIGURE 4.6 Axial and central chirality in coupled products via the Heck reaction.

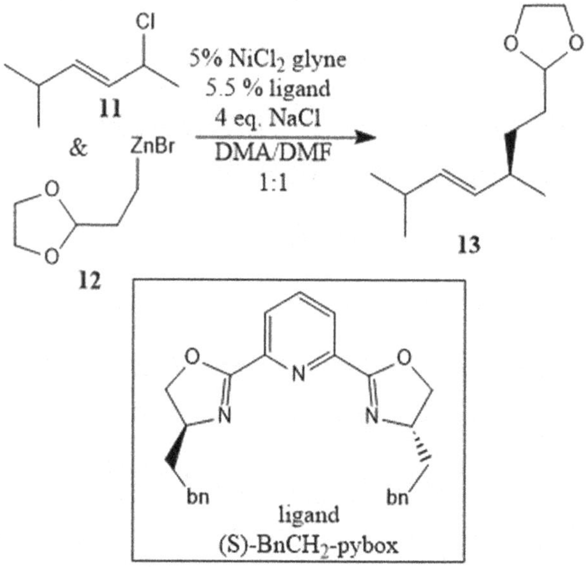

FIGURE 4.7 Negishi coupling using racemic allylic chlorides.

systems were used. This gave the products in very high yields with greater than 90% ee values.

One example with the Negishi coupling is Fu and Son[47] who developed a nickel-catalyzed asymmetric Negishi coupling to couple racemic secondary allylic chlorides (**11**) with primary alkyl zincs (**12**) (Figure 4.7). Yields ranged from 63% to 97% with %ee values in the range of 69%–98% using a Pybox ligand with the nickel catalyst system to give coupled systems such as the alcohol **13**.

CATALYST RECYCLING

For several reasons including but not limited to: cost, environmental protection, and safety of pharmaceutical products, the efficient removal (by way of separated isolation) of palladium-based catalysts is a goal. With methods that allow for not just recovery but reuse, important reaction materials like these catalyst systems can have second, third, or more "lives." One example applied to the Heck reaction uses palladium nanoparticle-engineered dibenzo-18-crown-6-ether/MCM-41 nanocomposite in water.[48] Use of this catalyst allowed a reduction in cost, high yield, an easier work-up, and shorter reaction time. Moreover, the nanocatalyst could be reused over seven runs with only slight loss in catalytic activity being observed and the catalyst was easily separated from the reaction mixture by filtration. Mukai and Yamada recently reviewed recycling catalysts in Suzuki couplings.[49] In addition to covering key principles, benefits, challenges, and applications of recyclable catalysts for green synthesis of bioactive molecules, they also discuss the potential for AI to advance green synthesis of pharmaceutical products.

BURKE AND THE USE OF MACHINE LEARNING TO STANDARDIZE AND AUTOMATE SYNTHESIS

In 2022, the Burke lab published a study that attempted to apply machine learning to synthetic chemistry. Specifically their goal was to identify optimum conditions for heterocyclic Suzuki–Miyaura reactions.[50] These attempts did not yield the desired standard protocols, a reality they attributed to reaction factors such as solvent and base being apparently nothing more than subjective preferences of synthetic chemists, rather than reasoned choices. Because the data set employed to "teach the machine" used thousands of examples from literature, it was arguably devoid of true logic that would allow such an identification to be made. Other machine learning endeavors had previously been more successful[51-54] including earlier work done by Burke.[51] Machine learning and AI are only going to get stronger and better, in all likelihood, increasing their utility in such endeavors. Even the aforementioned unsuccessful work by Burke could perhaps be made successful if the machine learning included data that included base properties so that the AI model could make correlations from what *appears to be* chaos. The potential for them to make work, like synthesis, more efficient is undeniably very high. Later that same year, in fact, Burke succeeded in identifying such an efficient protocol using machine learning.[55] To do this, they used their own data on a series of reactions where conditions were systematically changed, programming the results into the algorithm. This demonstrates the importance of quality data sets used in the programming or teaching of AI and other machine learning-based platforms. An important goal of the Burke lab is to further this goal of generating protocols that can be followed by an automated system for organic synthesis.[56][57]

USE OF ALTERNATIVE METALS

Nakamura et al.[58] and Fletcher et al.[59] have performed Suzuki–Miyaura couplings of racemic halides (**15**). Nakamura used iron as the catalyst to do enantioselective couplings of racemic alkyl bromides while Fletcher and his team used rhodium to catalyze the coupling of heterocycles with racemates. In Nakamura's case, after first mixing the iron metal with the chiral ligand briefly (Figure 4.8), the racemic halide and boronic ester are added. The reaction then proceeds very smoothly to give the coupled product in nearly quantitative yield with very good %ee values.

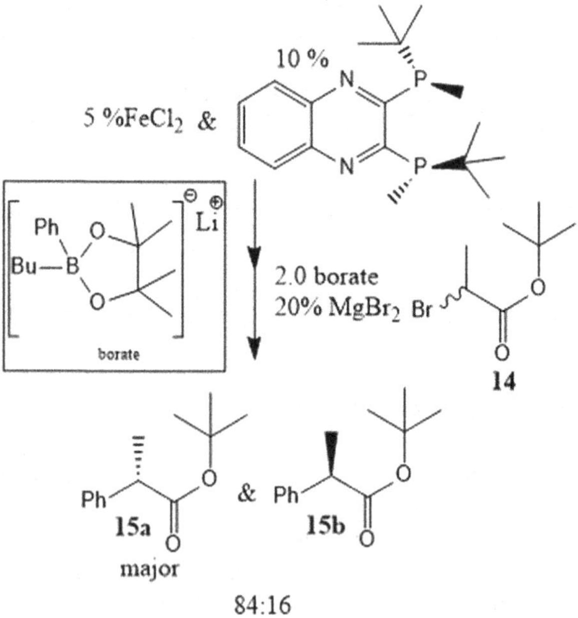

FIGURE 4.8   Nakamura's use of iron to perform Suzuki coupling.

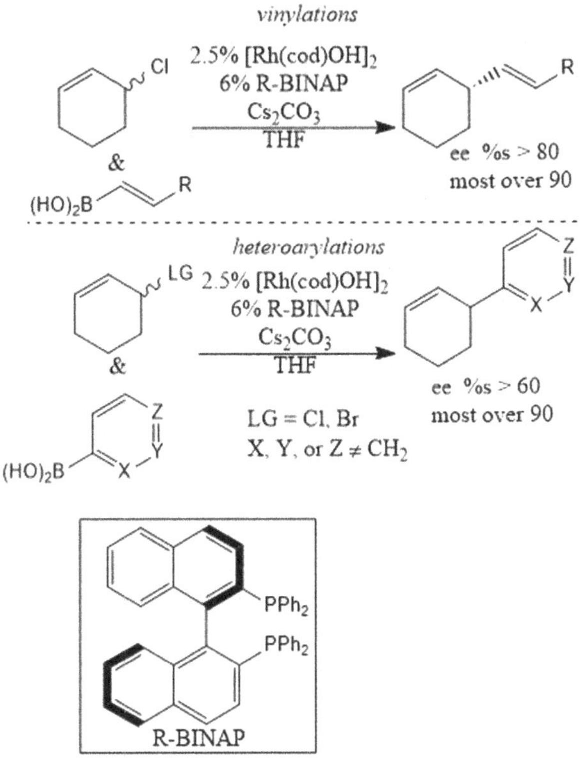

FIGURE 4.9   Fletcher's use of rhodium to perform Suzuki couplings.

The rhodium-catalyzed work of Fletcher (Figure 4.9) used R-BINAP as the chiral ligand to perform vinylations in yields ranging from 28% to 94% and %ee values ranging from 84% to 99%. In some cases, a higher yield was observed using Xyl-P-Phos as the ligand instead of BINAP. Meanwhile, the rhodium-catalyzed heteroarylations gave yield ranges of trace to 75% and %ee values ranging from 82% to 99%.

CHEMICAL TARGETS

Materials

Although an extensive coverage of these applications is beyond the scope of this series, it would be remiss to not mention that cross-coupling reactions have been used in the preparation of electronic, optical, electrochemical, and magnetic materials. Such uses have been reviewed.[60,61] Classes of compounds accessible include polythiophenes, polyfluorenes, polyphenylenes, polyheteroarylenes, oligo/polyarylethynylenes, and olig/polyarylvinylenes. These products have seen use as conducting polymers, semiconducting materials such as electron transport materials, hole transporting materials, and bipolar host materials; nonlinear optical materials; liquid crystals; light-emitting materials; light harvesting materials; electrochemical materials; and magnetic materials and high spin organic polyradicals.

Pharmaceutical Compounds and Natural Products

As evidenced in several reviews, even though all the varied palladium-catalyzed reactions certainly have used in the synthesis of natural products[62–66] and compounds of pharmaceutical interest,[67] and even the synthesis of agrochemicals such as herbicides, fungicides, and insecticides,[68] the examples presented in this chapter will be limited to the three Nobel-winning reactions. Interested readers are referred to the reviews cited earlier and countless others not cited herein for further examples of Nobel-winning and non-Nobel-winning palladium cross-coupling reactions.

One example of a *potential* pharmaceutical compound being made using cross-coupling in at least one step is GDC-6036, divarasib (**16**)[69] (Figure 4.10), a compound that is currently making its way through clinical trials for treating solid tumors with a KRAS G12C mutation including small-cell lung cancer, colorectal cancer, and others.[70] During the Phase 1 trial, a durable clinical response was observed across tumor types with the mutation and mostly low-grade adverse effects (side effects) were observed. This is a very promising result. To synthesize this promising compound, a Negishi coupling was used and the reaction proceeded atroposelectively. After the Negishi coupling furnished the new aryl–aryl bond, a subsequent series of steps furnished the target. The undesired atropisomer has the bond that was made rotated. Although single bonds are ordinarily able to freely rotate, the one made in this reaction, as many biphenyl derivatives do, has too high an energy barrier to freely rotate. The relationship between two such compounds that are "locked" in place is they are atropisomers of one another, a type of stereoisomer. The ligand used to achieve this level of selectivity was arrived at after a long series of trials and computational work described in the primary source and is beyond the scope of this discussion.

Another case, that of a drug target for asthma, PDE472 (**17**),[71] typifies how in drug discovery, the synthetic plan must be sometimes modified when

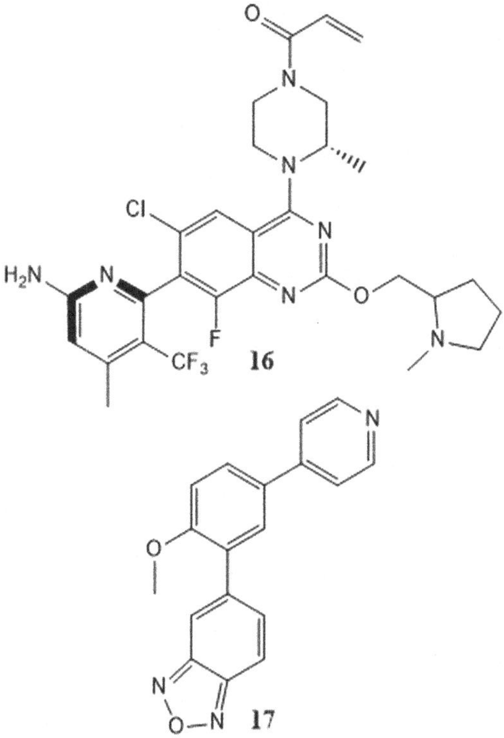

FIGURE 4.10   Interesting pharmaceutical compounds made using Suzuki coupling.

transitioning to the large-scale synthesis required for production (Figure 4.11). It likewise involves avoiding alternative methods (e.g., the tin-based Pd-catalyzed reactions—the Stille coupling) because of latent metal toxicity concerns for pharmaceutical products. All the routes to PDE472 in the discovery scale synthesis had yields in the mid 30% range. However problematic such low yields may be, it is not prohibitive in early stage drug development when the need does not exceed more than a few grams (if that much often much less is sufficient) to generate a lead. A different reality exists, however, when multi-kilogram scale syntheses are required, as when commencing clinical trials or bringing the drug to market; improvements were necessary if this promising compound were going to progress though the drug development pipeline. Such synthetic challenges undoubtedly have caused better-performing compounds to be abandoned by the pharmaceutical industry over the course of its history *in lieu* of less effective analogs that are easier to manufacture. Based on the previous experiments of the group developing this target, they chose to stick with the Negishi coupling conditions and ultimately increased the yield of the cross-coupling step to greater than 90%, finding that varying the addition

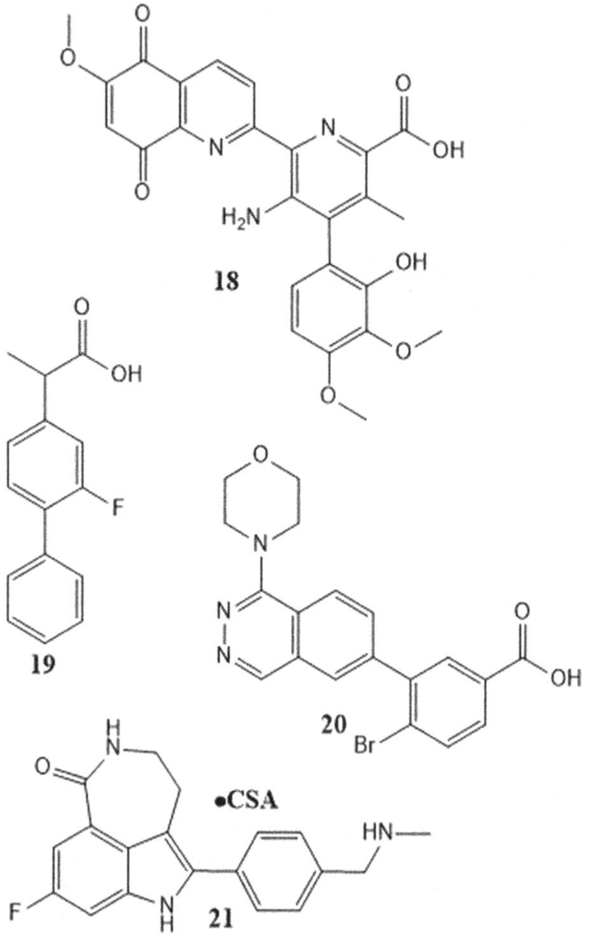

FIGURE 4.11   Interesting pharmaceutical compounds made using Negishi coupling.

times and additional stirring time of the ZnCl$_2$ has a profound impact on the reaction yield.

The antitumor antibiotic (±)-steptonogrin (**18**, Figure 4.10) (the ± indicates the product is a racemic mixture) was synthesized using a pair of Pd cross-coupling reactions—a Stille coupling, followed by a Suzuki–Miyaura coupling (Figure 4.12).[72] The first coupling takes advantage of the higher lability of the triflate leaving group, compared to the bromide, which is replaced in the second step. This pair of reactions set the stage by making two separate biaryl bonds. Additional chemical modification gave the final target.

Flurbiprofen (**19**) has been synthesized using a Suzuki coupling to forge the biphenyl bond (Figure 4.10).[73] This nonsteroidal anti-inflammatory drug

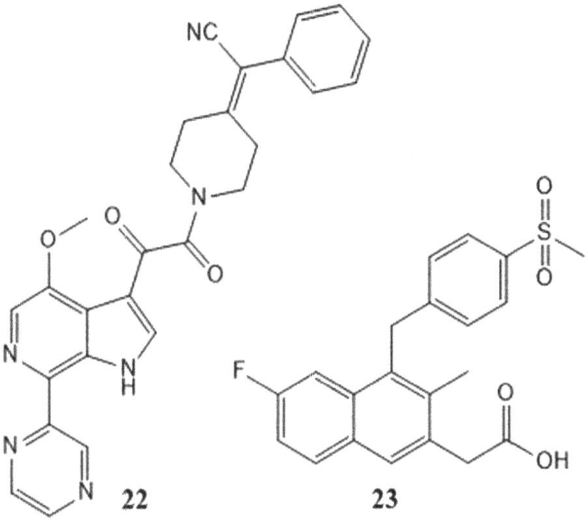

FIGURE 4.12   Other pharmaceutical compounds synthesized using Negishi coupling.

(NSAID) is used to alleviate arthritic pain.[74] Interestingly, in this synthesis, the leaving group is an amine (NH_2). Although an ibuprofen (trade name Advil) derivative, flurbiprofen is only available by prescription.

The Suzuki–Miyaura reaction was also employed in the synthesis of a compound (**20**)[75] (Figure 4.10) that positively regulates the production of pro-inflammatory mediators of TNFα and IL-1[76] which are involved in rheumatoid arthritis, Crohn's disease, and psoriasis. In the optimized protocol, the use of imidazole in the final step is believed to have aided in the removal of residual palladium from the previous step's Suzuki cross-coupling. This is beneficial and efficient because excessive care must be taken to remove all such heavy metal by-products before compounds can be consumed by humans.

Pf-01367338 (**21**)[77] was also synthesized using the Suzuki cross-coupling (Figure 4.10). This compound is believed to function as an inhibitor of an enzyme that repairs DNA damage in both cancerous and normal cells. This coupling joined a functionalized indole to benzaldehyde. Reductive amination of the aldehyde then gave the target. This compound (as the phosphonic acid salt) is known as Rucaparib today.

Similarly, the Negishi reaction has been used extensively in the synthesis of pharmaceutical compounds—for example, as one of the key steps in the synthesis of BMS-599793 (**22**), an HIV entry inhibitor.[78] Such inhibitors interrupt the life cycle of HIV by inhibiting the ability of the virus to enter the cell. The synthetic protocol (Figure 4.13) forges a bi-heteroaryl bond linking together a biaryl zinc and an azaindole. The biaryl zinc compound was made from successive treatment of the corresponding iodide with t-BuMgCl and

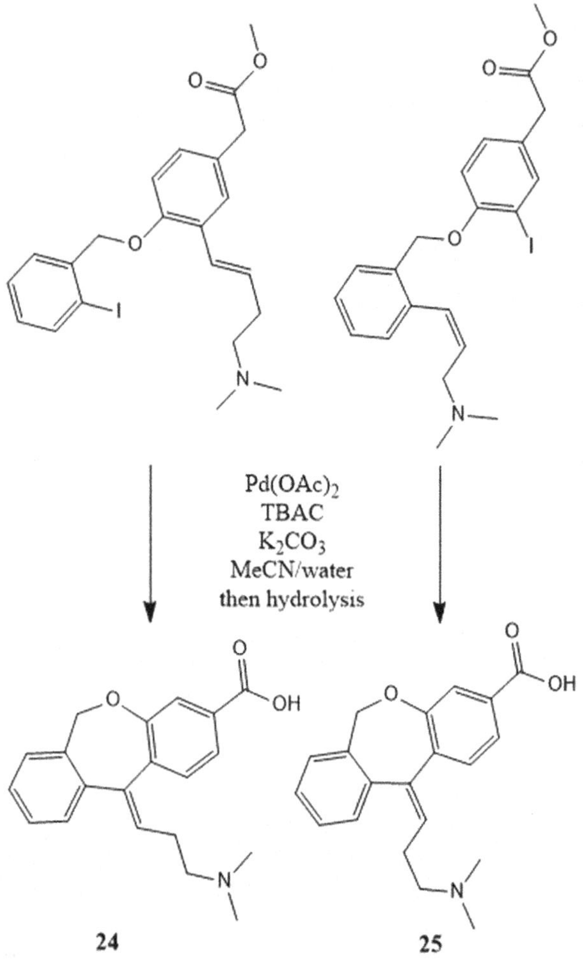

FIGURE 4.13   Olopatadine synthesized using the Heck reaction.

$ZnCl_2$, a necessary added step for any Negishi cross-coupling where the organozinc is not commercially available. This compound is also known as DS003 and is showing promise as a pre-(HIV) exposure prophylaxis candidate (Figure 4.12).[79]

Another example is during the preparation of a CHRT2 receptor agonist (**23**).[80] This receptor is believed to play a pro-inflammatory role in allergic properties. Due to this role, it is believed that such drug targets may hold a role in the treatment of asthma, atropic dermatitis, chronic obstructive pulmonary disease, and allergic rhinitis, among others. As seen earlier with PDE472, the small-scale medchem/discovery synthetic route needed modification upon

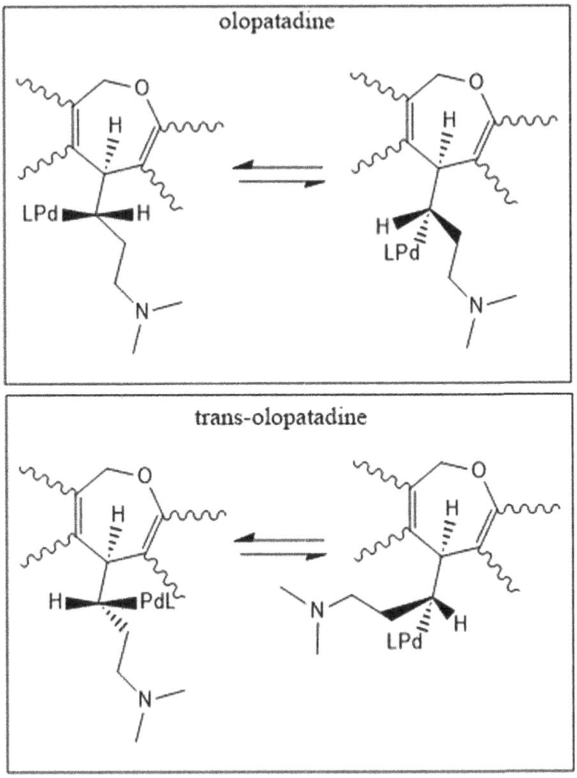

FIGURE 4.14   Intermediates in the synthesis of olapatadine and trans-olapatadine.

scaling up (Figure 4.15). One major difference is the purification method after the coupling. The discovery route used chromatography while scale-up used recrystallization; the former is not practical on large scales.

Another example is an intramolecular Heck reaction to give the antihistamine drug olopatadine (**24**).[81] The Heck reaction was used to make the oxepane ring in the target (Figure 4.16). It was found that which atom acted as the donor and which acts as the acceptor determines if the reaction progresses to the correct alkene configuration in olopatadine or its trans isomer (**25**). After the hydrolysis of the methyl ester that follows the Heck cyclization, the target is achieved. In both routes, a syn insertion is followed by a syn β-hydride elimination with the structure of the intermediates leading to the different alkene configurations (Figure 4.17).

The Heck reaction has similarly been used. Idebenone (**26**, Figure 4.18) is one such example that uses this reaction.[82] Originally designed as a treatment for Alzheimer's disease and Parkinson's disease, muscular illness, along with free-radical scavenger activities were also identified. Here (Figure 4.19),

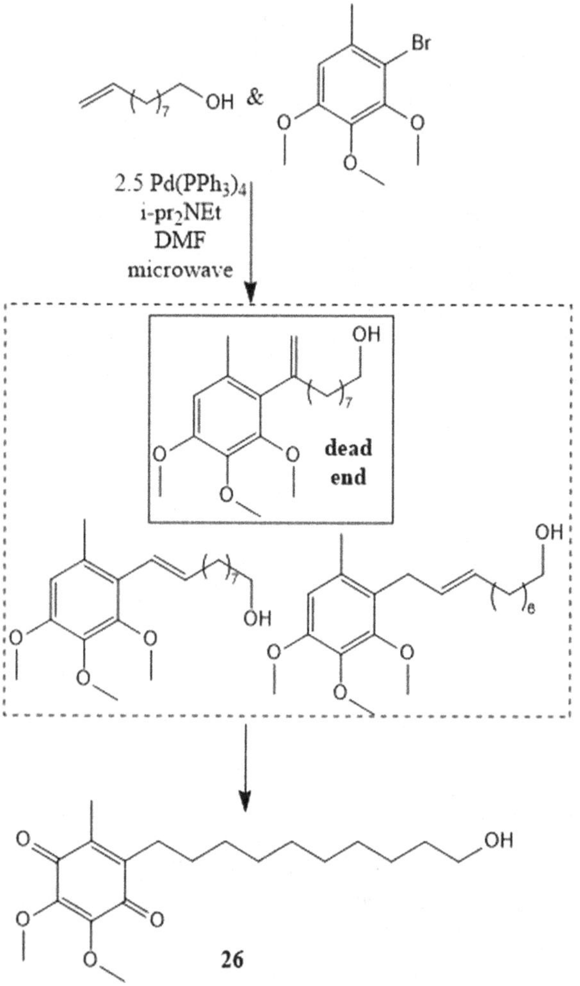

FIGURE 4.15   Syntesis of idebenone.

the Heck reaction was used to couple a long-chain alcohol to a substituted benzene ring. In addition to a small amount of the undesired regiochemical product where the wrong alkene carbon atom becomes attached to the benzene aryl ring, reductive elimination in the "wrong" direction was also observed. Although the former prevents progress to the target, the latter does not do so since the alkene in question is later hydrogenated.

These reactions (and other palladium-catalyzed cross-coupling reactions) have also been used in the synthesis of agrochemicals—chemicals used in

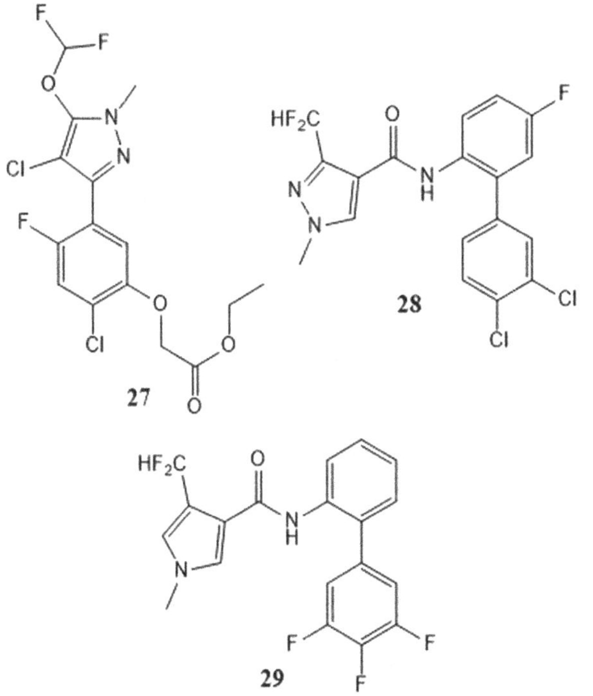

FIGURE 4.16   Agrochemicals synthesized by palladium-catalyzed reactions.

agriculture such as: fungicides, herbicides, and insecticides. One example is pyraflufen-ethyl (**27**), an herbicide that controls a wide range of annual broad-leaved weeds.[83–85] For this compound, a reaction was employed to affect a key modification of a highly substituted benzene ring (Figure 4.12). Here, the bromine-bearing position selectively reacted over the chlorine-bearing position.[8687]

The Suzuki reaction was used[88] to prepare bixafen (**28**) (Figure 4.12), a fungicide used to protect corn, soybeans, cereals, canola, peanuts, and pota-toes from diseases such as septoria leaf blotch that are resistant to other treatments.[89–91] After the coupling, additional modifications gave **28**.

A very similar target, fluxapyroxad (**29**), with similar modes of action against all major cereal diseases was also prepared. This polyfluorinated com-pound was synthesized using a Suzuki coupling[92–95] and also using a Negishi coupling (shown in Figure 4.12).[9697] In the latter, the second step hydrolyses the imide to reveal the free amine that is later modified to give the final target.

Palladium-catalyzed cross-coupling reactions have been used a myriad of times on the way to preparing natural products in the laboratory. Their use has

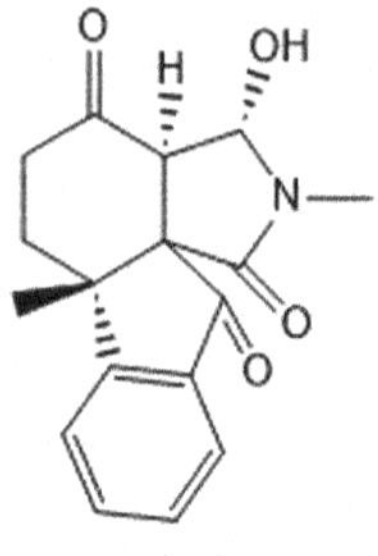

30

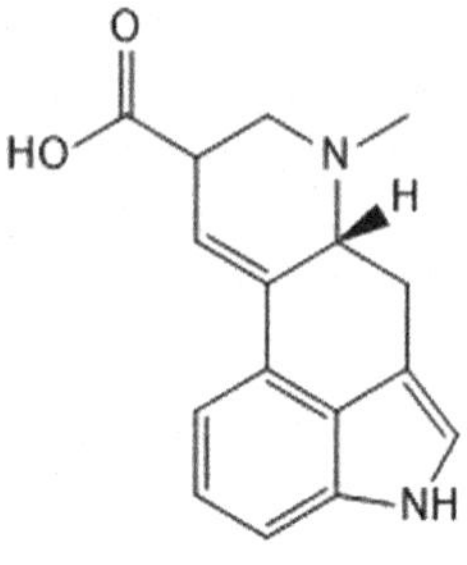

31

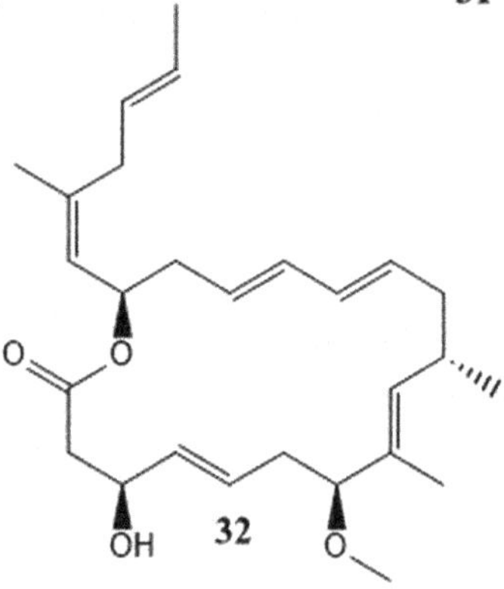

32

FIGURE 4.17 Natural products made using the Heck reaction.

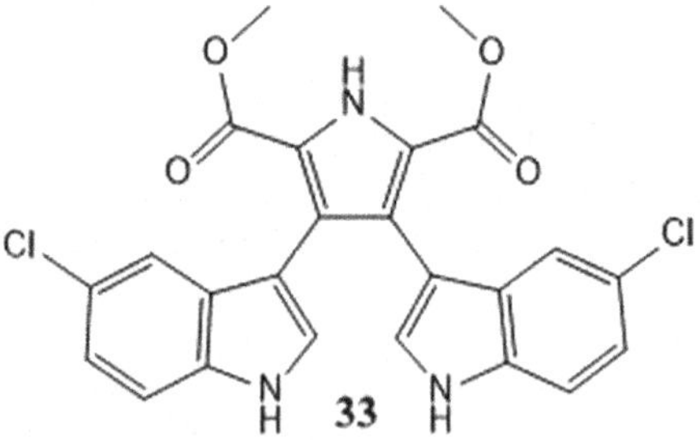

33

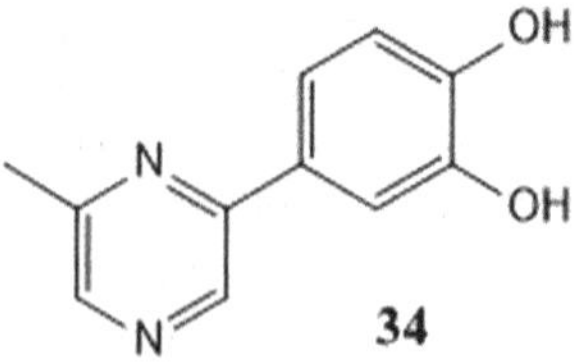

34

FIGURE 4.18 Natural products made using the Suzuki reaction.

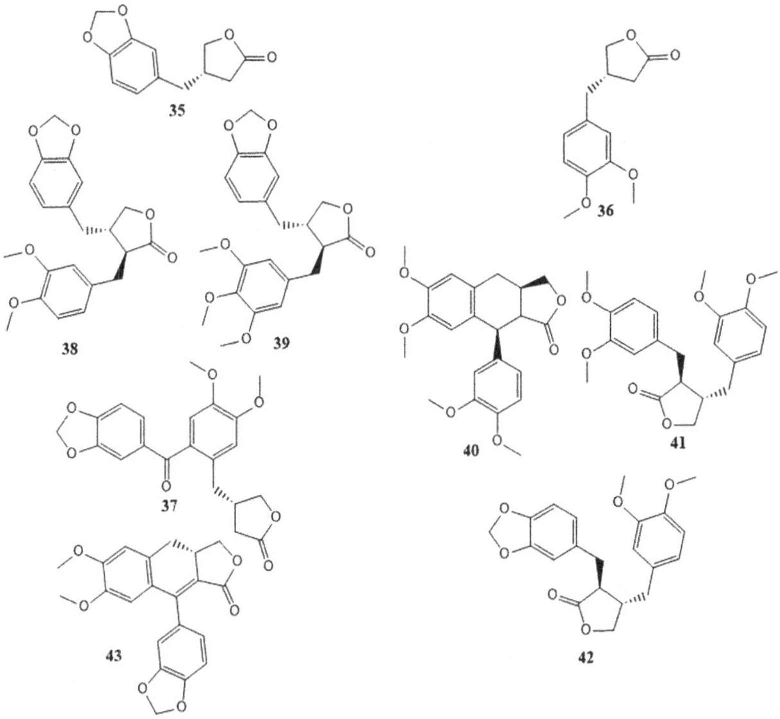

FIGURE 4.19 Family of natural products and their respective crucial intermediates prepared using Negishi coupling.

been reviewed many times as well, including but not limited to specifically the use of the reductive Heck reaction,[8] the intramolecular Heck reaction,[63] Heck macrocyclizations,[64] along with general cross-coupling reviews.[65] Owing to the effectively limitless flexibility in the type of carbon–carbon single bond that can be made using these methods, thousands of syntheses have relied on them for one or more steps along a multistep pathway. The several cases covered herein do the utility and popularity of these reactions and the other palladium-catalyzed couplings no justice.

One example that employs a reductive Heck reaction furnishes the tetra-cyclic natural product exiguaquinol (**30**).[98] Here (Figure 4.13), the reductive Heck reaction was used to generate one of the five-membered rings in the target. Additional steps gave access to the natural product target.

In another case, an intramolecular Heck reaction was used to complete the synthesis of (+)-lyseric acid (**31**).[99] The cross-coupling method was used to complete an important ring closure ahead of final steps to finish the synthesis. During their studies, Jia and co-workers found that the use of Ag_2CO_3 as the base was necessary as it not only acted as a base, but also a halide scavenger to suppress isomerization of the alkene during the Heck reaction.

The Heck reaction was used in the Goswami[100] and Maier[101] syntheses (Figure 4.13) of biselyngbyolide B (**32**), the aglycone of the novel anticancer macrolide biselyngbyaside. This compound was isolated from a cyanobacterium[102] and was found to induce apoptosis in osteoclasts.[103] Later, it was discovered that the aglycone was even more potent at inducing apoptosis and was a strong endoplasmic reticulum stress inducer.[104]

The Suzuki coupling has also been used in the synthesis of natural products, including what was the first total synthesis of the bisindole lynamycin D (**33**).[105] After the one-pot bis Suzuki coupling (Figure 4.15), deprotection gave lynamicin D in very good yield. This compound was isolated from multiple marine actinomycetes and showed activity against drug-resistant strains of *Staphylococcus aureus* and *Enterococcus faecium*.[106][107]

Aspongopus chinensis Dallas is an insect found in China that has been used as a food and traditional medicine. It was found that it exhibits biological activities such as anticancer, analgesic, and angiogenesis, among others. One compound isolated from insect *A. chinensis* is aspongpyrazine A (**34**).[108] Bendre and co-workers[109] completed a total synthesis of this compound by employing palladium nanoparticles prepared using palladium chloride and oxytocin. Subsequent deprotection of the methyl ethers gave the target.

The Negishi coupling has likewise been used many times. For example, the Negishi coupling was employed to build common intermediates (**35–37**) en route to several compounds (Figure 4.19). One intermediate (**36**) led to two targets: (±)-busehernin (**38**) and (±)-yatein (**39**). A second intermediate (**35**) led to three targets: (±)-dimethylretrodendrin (**40**), (±)-dimethylmetairesinol (**41**), and (±)-kusonokinin (**42**). The third intermediate (**37**) led to (±)-collinusin (**43**).[110]

Another triumph of the Negishi cross-coupling reaction is the synthesis (Figure 4.20) of the paracyclophane (−)-cylindrocyclophane F (**44**).[111] This compound was isolated from blue-green algae and was one of multiple 22-member macrocycles found to have cytotoxic potential against tumor cell lines.[112–114] This family of compounds were the first naturally occurring paracyclophanes.

Finally, a Negishi coupling was used as one of the key steps in the Smith Group's synthesis (Figure 4.20) of discodermolide (**45**).[115][116] This natural product is an important microtubule-stabilizing agent, making it potentially useful in the battle against cancer. This protocol joins together two fragments with many stereocenters and required additional modification in the way of both further elaboration and deprotections to arrive at the final target.

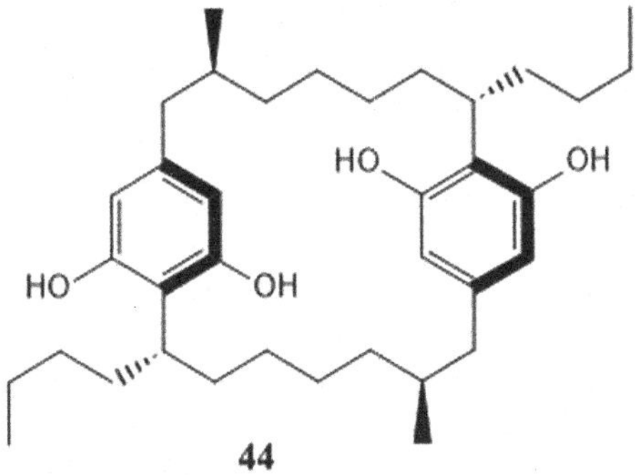

44

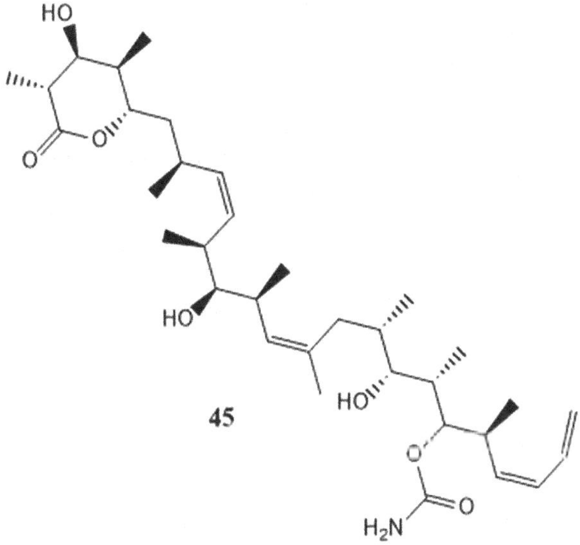

45

FIGURE 4.20   Additional synthetic targets reached using Negishi coupling.

SUMMARY

The reactions covered in this chapter have transformed the field of organic synthesis dramatically. They have done so by making possible or more efficient the formation of effectively any carbon–carbon single bond. That these methods (and others not awarded a Nobel) can also be done in a stereoselective manner only makes them more powerful. Each of them have been used in at least one of the synthetic steps to prepare many important molecules.

GENERAL REFERENCES

Akira Suzuki. Biography. www.nobelprize.org/prizes/chemistry/2010/suzuki/biographical/, Last accessed 8/21/24.

Akira Suzuki. Nobel Prize Lecture. www.nobelprize.org/uploads/2018/06/suzuki_lecture.pdf, Last accessed 8/21/24.

Ei-ichi Negishi. Biographical. www.nobelprize.org/prizes/chemistry/2010/negishi/biographical/, Last accessed 8/21/24.

Ei-Ichi Negishi. Nobel Prize Lecture. www.nobelprize.org/uploads/2018/06/negishi_lecture.pdf, Last accessed 8/21/24.

Laszlo Kürti and Barbara Czakó, *Strategic Applications of Named Reactions in Organic Synthesis*, Elsevier Academic Press, 2005 (for reaction mechanisms).

Luke R. Odell and Mats Larhed, Richard F. Heck. Biographical. www.nobelprize.org/prizes/chemistry/2010/heck/biographical/, Last accessed 8/21/24.

Scientifc Background on the Nobel Prize in Chemistry 2010. palladium-catalyzed cross couplings in organic synthesis www.nobelprize.org/uploads/2018/06/advanced-chemistryprize2010-1.pdf, Last accessed 8/21/24.

REFERENCES

(1)   *Journal of the American Chemical Society* **1961**, 4023.
(2)   *Journal of the American Chemical Society* **1968**, 5538.
(3)   *Journal of the American Chemical Society* **1968**, 5518.
(4)   *Journal of the American Chemical Society* **1968**, 5526.
(5)   *Journal of the American Chemical Society* **1969**, 6707.
(6)   *Journal of Organic Chemistry* **1972**, 2320.
(7)   *Journal of Organic Chemistry* **1983**, 4236.
(8)   *Chemistry Select Reviews* **2019**, 2057.
(9)   *Journal of the American Chemical Society* **1941**, 2316.
(10) *Journal of Organic Chemistry* **1952**, 1630.
(11) *Journal of the American Chemical Society* **1971**, 1487.
(12) *Chemical Communications* **1972**, 144.
(13) *Journal of the American Chemical Society* **1972**, 4374.
(14) *Journal of Organometallic Chemistry* **1975**, C39.
(15) *Chemical Communications* **1976**, 596.
(16) *Journal of the American Chemical Society* **1976**, 6729.
(17) *Journal of Organic Chemistry* **1977**, 1821.
(18) *Chemical Communications* **1977**, 683.

(19) *Chemical Communications* **1973**, 606.
(20) *Chemical Communications* **1973**, 874.
(21) *Tetrahedron Letters* **1977**, 411.
(22) *Journal of Organometallic Chemistry* **2002**, 23.
(23) *Journal of the Chemical Society Chemical Communications* **1972**, 144.
(24) *Journal of the Chemical Society Chemical Communications* **1976**, 596.
(25) *Accounts of Chemical Research* **1982**, 340.
(26) In *Aspects of Mechanism and Organometallic Chemistry*; Brewster, J. H. Ed., Plenum Press: New York, 1978, p. 285.
(27) *Chemical Letters* **1997**, 301.
(28) *Tetrahedron Letters* **1979**, 3437.
(29) *Journal of the Chemical Society Chemical Communications* **1979**, 866.
(30) *Synthetic Communications* **1981**, 513.
(31) *Synlett* **1992**, 207.
(32) *Acta Chimica Scandinavica* **1973**, 521.
(33) *Molecules* **2020**, 5506.
(34) *Journal of Organic Chemistry* **2003**, 5660.
(35) *Angewandte Chemie International Edition* **2003**, 4856.
(36) *Angewandte Chemie International Edition* **2003**, 1407.
(37) *Journal of Organic Chemistry* **2005**, 161.
(38) *Journal of Organic Chemistry* **2008**, 3854.
(39) *Tetrahedron Letters* **2006**, 1909.
(40) *Synlett* **2006**, 2953.
(41) *Nature Reviews Chemistry* **2020**, 584.
(42) *Current Organic Chemistry* **2004**, 781.
(43) *Advanced Synthesis & Catalysis* **2007**, 1533.
(44) *Chemical Reviews* **2015**, 9587.
(45) *Journal of the American Chemical Society* **2022**, 15026.
(46) *Journal of the American Chemical Society* **2018**, 11067.
(47) *Journal of the American Chemical Society* **2008**, 2756.
(48) *Applied Organometallic Chemistry* **2018**, e4271.
(49) *Knowledge* **2023**, 1.
(50) *Journal of the American Chemical Society* **2022**, 4819.
(51) *Angewandte Chemie International Edition* **2019**, 4515.
(52) *Angewandte Chemie International Edition* **2020**, 13253.
(53) *Chemical Science* **2021**, 2931.
(54) *Nature Communications* **2020**, 3878.
(55) *Science* **2022**, 399.
(56) *Angewandte Chemie International Edition* **2018**, 4192.
(57) *Nature* **2022**, 92.
(58) *Chemical Communications* **2019**, 1128.
(59) *Nature Communications* **2017**, 15762.
(60) *Science and Technology of Advanced Materials* **2014**, 044201.
(61) *Organic Chemistry Frontiers* **2015**, 416.
(62) *Angewandte Chemie International Edition* **2005**, 4442.
(63) *European Journal of Organic Chemistry* **2021**, 2057.

(64) *Natural Products Reports* **2021**, 1109.
(65) *Molecular Diversity* **2022**, 647.
(66) *Indo Global Journal of Pharmaceutical Sciences* **2012**, 351.
(67) *Journal of the Brazilian Chemical Society* **2014**, 2186.
(68) *Journal of Agricultural and Food Chemistry* **2018**, 8914.
(69) *Journal of the American Chemical Society* **2022**, 20955.
(70) *New England Journal of Medicine* **2023**, 710.
(71) *Organic Processes Research and Development* **2003**, 436.
(72) *Journal of Organic Chemistry* **2013**, 12338.
(73) *Chinese Chemical Letters* **2006**, 461.
(74) Flurbiprofen (oral route). www.mayoclinic.org/drugs-supplements/flurbipro fen-oral-route/description/drg-20069793, 8/18/24
(75) *Synthetic Letters* **2012**, 1564.
(76) *Journal of Medicinal Chemistry* **2008**, 627.
(77) *Organic Processes Research and Development* **2012**, 1897.
(78) *Organic Processes Research and Development* **2013**, 907.
(79) *AIDS* **2021**, 1907.
(80) *Organic Processes Research and Development* **2013**, 651.
(81) *Journal of Organic Chemistry* **2012**, 6340.
(82) *Organic Processes Research and Development* **2011**, 673.
(83) *Yuki Gosei Kagaku Kyokaishi* **2003**, 2.
(84) *Pub source*; assignee: country, 1990; Vol. version number, p pages.
(85) *Brighton Crop Protection Conference WEeds, British Crop Protection Council* **1993**, 35.
(86) *Japanese Kokai Tokkyo* **2010**, JP2017206453A.
(87) Shuhei Kubota. Manufacturing method of pyrazole derivative and intermediate products thereof. **2017**, JP2017206453A.
(88) *PCT International Applications* **2009**, WO2009135598A1.
(89) *Pesticide Biochemistry Physiology* **2012**, 171.
(90) *Julius-Kühn-Archiv* **2010**, 142.
(91) *Bioactive Heterocyclic Compound Classes: Agrochemicals*; Wiley VCH Verlag GMBH and Co., 2012.
(92) *European Journal of Organic Chemistry* **2017**, 6566.
(93) *Farming Zhuanli Shenquig* **2015**, CN104529786A.
(94) *PCT International Applications* **2009**, WO2009156359A2.
(95) Nantong Jiahe Chemical Co Ltd. East China University of Science and Technology , CN107382734A. The Suzuki coupling reactions of the amino palladium catalyst nitro-chlorobenzene of dichloro two. **2017**.
(96) *Bioorganic & Medical Chemistry* **2016**, 317.
(97) *PCT International Applications* **2010**, WO2010094736A1.
(98) *Organic Reactions* **2002**, 157.
(99) *Journal of Organic Chemistry* **2013**, 10885.
(100) *Organic Letters* **2016**, 1908.
(101) *Journal of Organic Chemistry* **2018**, 4554.
(102) *Organic Letters* **2009**, 2421.
(103) *Journal of Cellular Biochemistry* **2012**, 440.

(104) *Chemistry Letters* **2014**, 287.
(105) *Bioorganic & Medical Chemistry* **2017**, 1622.
(106) *Journal of Natural Products* **2008**, 1732.
(107) *Interdisciplinary Sciences: Computational Life Sciences* **2014**, 187.
(108) *RSC Advances* **2015**, 70985.
(109) *Journal of Organometallic Chemistry* **2020**, 121093.
(110) *Journal of Organic Chemistry* **2019**, 2920.
(111) *Chemistry: A European Journal* **2018**, 16770.
(112) *Journal of the American Chemical Society* **1990**, 4061.
(113) *Tetrahedron* **1992**, 3001.
(114) *Nature Chemical Biology* **2017**, 916.
(115) *Organic Letters* **1999**, 1823.
(116) *ChemBioChem* **2000**, 171.

Index

A

Acyclic diene metathesis (ADMET), 90
Advinus Therapeutics, 64
Al Gore, 22
Alkene metathesis, 78–80
Alternative metals, 110–111
American education system, 23
Amphidinolide P product, 91–92
Artificial intelligence (AI), 99
Asymmetric epoxidation catalysts, 62
Asymmetric hydrogenations, 67–71
Auckland Cancer Research Centre, 64
Aung San Suu Kyi, 5
Automate synthesis, 109–110
Avogadro law, 18
Axial and central chirality, 108

B

Banerjee, A., 9
Benzazepines, 92
Benzazocines, 92
Bergström, S., 9
Berkessel's ligand for syn epoxidations, 63
Bertozzi, C., 15–16
Bipolar host materials, 112
Bohr, A., 9
Bragg, W. L., 9
Breslow, D., 100
Burke, M., 34–35
    Burke lab, 109–110
Burnell, J. B., 22
Butenandt, A., 5

C

Carbon–carbon bonds, 99–100
Carlson, R., 14, 22
Carver, G. W., 19
Carvone, 54, 55
Catalyst recycling, 109
C–H activation, 34
Chauvin, Y., 47, 77, 79–81
Chiral additives, asymmetric dihydroxylations, 61

Chiral morpholines, 73
Chiral phosphine ligands, 68, 69
CHRT2 receptor agonist, 116
Cinderella Stories, 32
Continuous flow/flow reactors, 86–88
Cori, G., 9
Cornforth, J. W., 53
Cram, D. (wrong number phone call), 15
Cross-coupling reactions, 96
Cross metathesis, 90
Curie, M. and the Nobel prize, 20–22
Curie, P., 6

D

de Hevesy, G., 14–15
Diastereomers, 54, 55
Diels–Alder reaction, 97
Diethyl tartrate (DET) additive, 58
Diphosphines, 69
Domagk, G., 5
Drugs for Neglected Disease Institute, 64
Duflo, E., 9

E

Economic Sciences prize, 1
Electrochemical materials, 112
Electron transport materials, 112
Enantiomers, 54–57
Enantioselective palladium-catalyzed
    cross–cross-coupling reactions, 107–108
Epoxidations of geraniol, 60
ERATO molecular catalysis, 71
Exiguaquinol (tetracyclic natural product), 121

F

Fermi, E., 13
Fibiger, J., 13
Fletcher's use of rhodium, Suzuki coupling, 111
Flow reactors, 85, 106–107
Fluorinated compounds, 34
Flurbiprofen, 114–115
Franck, J., 14

Franklin, R., 21–22
Friedel–Crafts reactions, 90
Fungicides, 119

G
Gajdusek, D. C., 6
Goodenough, J. B., 36
Grignard cross-couplings, 103
Grignard reaction, 97
Grubbs, R. H., 47–48, 77
    catalysts, 84
    work, 83–85

H
Haber, F., 6
Hahn, O., 13
Hammarskjöl, D., 4
"Handedness" of molecules, 54, 55
Hawking, S., 19
Heck, R., 49, 96–123
    The Heck reaction mechanism, 97,
        100–102
Henry, P., 100
Henry, W., 9
Herbicides, 119
High spin organic polyradicals, 112
Hole transporting materials, 112
Hydrizogenation of unprotected indoles, 74

I
Idebenone synthesis, 117, 120
Ig®Nobel Prize, 40–42
Insecticides, 119
Ions of palladium (II), 96
    carbon–carbon bonds, 99–100
    catalyzed coupling reactions, 96
    nobel-winning cross-coupling
        reactions, 97–98

J
Jean-Paul, S., 5
Joliot-Curie, I., 6, 9

K
Karlfeldt, E. A., 4
Knowles, W., 45–46, 53, 67–71
Kornberg, A., 9
Kornberg, R., 9
Kuhn, R., 5
The Kumada reaction, 99
Kuru (role of prions), 6

L
Labouisse, H. R., 9
Lactams, 92, 93
L-DOPA synthesis, 70
Leadbeater lab, 106–107
Le Châtelier, H. L., 19
Le Duc Tho, 5
Lewis, G., 18
Light-emitting materials, 112
Light harvesting materials, 112
Liquid crystals, 112
(+)-Lyseric acid synthesis, 121

M
Macrocyclizations, 121
Madness, M., 32
Magnetic materials, 112
Matsugi's recyclable fluorescent metathesis
    catalysts, 86
May-Britt, M., 9
Meitner, L., 21
Mendeleev, D., 19
Menthol synthesis, 72
the Merchant of death (Nobel, Alfred), 2
Merrifield, R. B., 45, 85
Metathesis reactions
    alkene, 78–79
    Chauvin's work, 79–81
    definition, 77
    general, 78
    Grubbs's work, 83–85
    materials, 88–90
    polymerization, 79
    ROMP, 79
    Schrock's work, 81–83
Microwave heating, 106–107
Mohammadi, N., 5
Monic, A. E., 14
Morello, T., 15–16
Moser, E. I., 9
Müller, P. H., 14
Mydral, A., 9
Mydral, G., 9

N
Nakamura's use of iron, Suzuki coupling, 110
Natural product synthesis, 65–67, 90–93,
    112–123
Negishi, E., 49–50
    cross-coupling, 97, 98, 102–104, 108,
        122, 123

N-Morpholine oxide (NMO), 61
Nobel Families, 6
Nobel prize (Nobel laureates)
  achievements, 3
  ancient Greece, 2
  cash prize, ornate diploma and medal, 2
  chemistry prize, 33–34
  controversies, 16–17
  disproven works, 13
  Foundation's restricted funds, 3
  future of, 32–33
  of misconduct, 13–14
  modern science labs, 4
  Nazis and disappearing Nobel Prizes, 14–15
  Nobel Snubs, 17–20
  permutations, 3
  selection of, 10–13
  and society, 39–40
  wrong number phone call (Cram, Donald), 15
Nobel Prize controversies, 16–17
Nobel Prize-winning families, 6–9
Nocera, D., 35
Non-Caucasian awardees, 22
Nonlinear optical materials, 112
Nonsteroidal anti-inflammatory drug (NSAID), 114–115
Noyori, R., 46, 53–74
Nuclear magnetic resonance (NMR), 57

O
Olefin metathesis, 97
Oligo/polyarylethynylenes, 112
Olig/polyarylvinylenes, 112
Olopatadine (antihistamine drug), 117

P
Pääbo, S., 9
Palladium-catalyzed coupling reactions, 96
Palladium orbital energies, 96–97
Pasternak, B., 5
The Peace Prize, 16
Pf-01367338 reaction, 115
Pharmaceutical compounds, 112–123
Physiology or Medicine award, 1
Polyfluorenes, 112
Polyheteroarylenes, 112
Polymer-supported catalysts, 86
Polyphenylenes, 112
Polythiophenes, 112
Pregl, F., 44

Prelog, V., 53
Proton NMR (HNMR), 57

R
Race and the Nobel Prize, 22–31
Racemic allylic chlorides, 108
Reductive Heck reaction, 102
Regiochemistry, 58
Ring-Closing Metathesis (RCM) reactions, 77
Ring-Opening Metathesis Polymerization (ROMP), 79
Robinson, R., 44
Roosevelt, E., 22

S
S and R BINAP, structure of, 70, 71
Sanford, M., 34
Sanger, F., 33
Santa Barbara Amorphous (SBA-15), 86
Schrock, R., 48–49, 77
  initial metathesis catalysts, 82
  work, 81–83
Sharpless, K. B., 33, 46–47, 57–60
  asymmetric aminohydroxylation, 67
  SAD, 59–62
  SAE, 58, 59
Siegbahn, K., 9
The Sonogashira reaction, 99
Steinman, R., 4
Stereochemistry, 58
Stereoisomers, 56–57
The Stille reaction, 99
Subdisciplines of chemistry, Nobel Prize, 37–38
Suzuki, A., 50, 96–123
  cross-coupling, 104–105, 110, 111
The Suzuki–Miyaura reaction, 115
Sveriges Riksbank Prize in Economic Sciences, 1
Swedish Academy of Sciences, 1
Swift, T., 36

T
Tandem processes, 72–73
Thalidomide, 54, 55
Thomson, G., 9
Tiananmen Square in 1989, 5
Time Magazine in 1941, 19
Tinbergen, N., 10
Trans-olopatadine formation, 120

V
Vietnam Peace Accord, 5
von Baeyer, A., 44
von Euler-Chelpin, H., 9
von Euler, U., 9
von Laue, M., 14
von Ossietzky, C., 5

W
Wacker Process, 100
Water-compatible systems, 72
White, M. C., 34

Whittingham, M. S., 36
Wilkinson's catalyst, 67
Wittig reaction, 97
Women and the Nobel prize,
　　20–22
Woodward, R. B., 44

X
Xiaobo, L., 5

Y
Yoshino, A., 36

Printed in the United States
by Baker & Taylor Publisher Services